Data Center Networking

Deke Guo

Data Center Networking

Network Topologies and Traffic Management
in Large-Scale Data Centers

Deke Guo
School of System Engineering
National University of Defense Technology
Changsha, Hunan, China

ISBN 978-981-16-9370-0   ISBN 978-981-16-9368-7   (eBook)
https://doi.org/10.1007/978-981-16-9368-7

Jointly published with Tsinghua University Press
The print edition is not for sale in China (Mainland). Customers from China (Mainland) please order the print book from: Tsinghua University Press.

This Springer imprint is published by the registered company Springer Nature Singapore Pte Ltd.
The registered company address is: 152 Beach Road, #21-01/04 Gateway East, Singapore 189721, Singapore

Preface

Background

The data center aims to interconnect the hardware resources, such as large-scale servers and network facilities, with dedicated topologies. Thus, data center acts as a massive resource pool, including computing, storage, network and other resources, to provide elastic services for various applications.

Data center networking (DCN) plays an essential role for the design of data centers. DCN aims to connect all switches and servers together using an outstanding topology, so as to achieve better network performance. Many applications hosted by data centers cause intensive data interaction across multiple servers. DCN becomes the bottleneck of data centers, and affects user's experience directly. Therefore, it is urgent to bring more innovations into the field of DCN, so as to promote the development of data centers and its applications.

The research about DCN has become a highlight field in the academia as well as the industry communities. There still exist a series of fundamental challenges for speeding the development of the data centers. Specifically, this book focuses on two challenges, including the topology design and the collaborative management of traffic inside each data center.

This book deeply discusses the design and optimization methods of several novel network topologies for data centers, so as to realize the high bandwidth, high fault tolerance and high scalability. Furthermore, this book reports the cooperative transmission mechanisms of several correlated traffic patterns in a data center, which can significantly reduce the bandwidth consumption. The majority content in the book come from our recent papers published in journals or conferences, and systematically report representative research results in the field of data center network.

Content Organization

The book consists of ten chapters, which can be divided into three parts. Part I presents the background and overview of data centers and then discusses the state-of-the-art network topologies of data centers, in Chaps. 1 and 2, respectively.

Chapter 1 reports the evolution trend of data centers, illustrates the application fields of data centers, and summarizes the important challenges of data center networking.

Chapter 2 proposes a new taxonomy that divides current data center topologies into five categories, including the switch-centric, server-centric, modular, random, and wireless topologies.

Part II introduces our four network topologies of data centers in Chaps. 3–6. They are the server-centric network topology HCN, modular network topology DCube, hybrid network topology R3, and wireless network topology VLCcube.

Chapter 3 introduces two server-centric network topologies, HCN and BCN, based on the compound graph theory. They are constructed using homogeneous servers, each of which just has two NICs, and commodity switches.

Chapter 4 introduces a family of intra-module network topologies, called DCube. Each DCube interconnects a large number of dual-port servers and low-end switches. Many DCube modules can further form a new modular data center.

Chapter 5 introduces a general design methodology for DCN topologies using the compound graph theory, and discusses a hybrid network topology R3, which is the compound graph of a structured topology and a random topology.

Chapter 6 proposes a hybrid network topology, VLCcube, by introducing extra visible light communication (VLC) links into data centers. Specifically, VLCcube augments Fat-Tree, a representative DCN in production data centers.

Part III focuses on the cooperative transmission of correlated flows inside a data center. Managing those correlated flows is very essential for efficiently utilizing the network resource and speeding the data applications.

Chapter 7 introduces the cooperative transmission management problem of Incast. The basic idea is to perform the in-network data aggregation, during the transmission phase as early as possible rather than just at the receiver side.

Chapter 8 considers how to realize the in-network aggregation of correlated Shuffle transfer, so that the bandwidth consumption can be considerably reduced. This chapter also introduces the scalable traffic forwarding mode based on Bloom filters.

Chapter 9 makes the first step towards the study of in-network aggregation of uncertain Incast transfer.

Chapter 10 introduces the cooperative traffic management of multicast. This chapter proposes the minimum cost forest (MCF) model for such a multicast transfer with uncertain senders.

The author's research work has been partially supported by the National Natural Science Foundation of China under Grant No. U19B2024.

Changsha, China   Deke Guo
September 2021

Contents

Part I
Basic Knowledge

Chapter 1
Introduction of Data Center

Abstract This chapter presents the background and overview of the data center from the three dimensions. First, we introduce the basic concept and history of data centers, especially the evolution of data centers in the era of cloud computing, the internet of things and big data. Next, we illustrate the potential application fields of data centers, such as networked computing, networked storage, the data analysis. In the end, we summarize representative challenges of data center networking, including the functionality customization, network virtualization, the scalable network topology, traffic management, et al.

1.1 Evolution of Data Centers

1.1.1 Basic Concept and Intuitive Classification

The data center is a complex system to interconnect hardware resources, such as large-scale servers and network facilities, with dedicated topologies. It also contains redundant or backup power supplies, redundant data communication connections, environmental controls and various security devices. By doing so, the data center usually acts as a massive resource pool, including computing, storage, network and other resources, to provide elastic services to all applications. New computing paradigms and applications, such as cloud computing, internet of things and big data analysis, further accelerate the development of modern data centers. As a result, the data center is regarded as a basic information infrastructure for countries and IT enterprises, and plays an indispensable role in the economic field, the science field and technology field, and daily life.

The emergence of data centers aims to meet the strong demand for efficient organization and management of massive data [1]. With the growing customers, the banking and financial industry has urgent demand for efficient data storage and management technologies in the early days. The boom of the Internet and the revolution of digital society significantly speed the increase of available data in Internet. Consequently, a variety of network storage architectures have been proposed to realize the storage and management of those massive data. Mainstream architectures include the direct attached storage (DAS), the network attached storage (NAS) and the storage area networks (SAN).

A typical DAS system is made of data storage devices directly attached to the computer or a server accessing it via SCSI cables or fiber channels, as opposed to the storage accessed over a computer network. With SCSI cables, a server can accommodate up to 16 storage devices, while this number is 126 when optical fibers are utilized. However, DAS falls short of the low scalability, and high maintenance complexity. First, the storage hardware prefers to serve the attached server, thus cannot be shared and utilized by other servers. Whenever the storage resource of any server is always overloaded, additional storage devices have to be attached, even the storage resources at other servers are under-utilized and wasted. Moreover, servers may support diverse applications, which differ in the number of consumed storage resources. This phenomenon further lowers the resource utilization of the DAS system. DAS usually offers low storage capacity at the cost of a few servers, and hence is suitable for small-scale or medium-scale systems. Generally, DAS fails to meet the increasing requirements of current storage applications in terms of the large capacity, high reliability, high availability, high performance, dynamic scalability, easy maintenance, et al. The key insight of filling this gap is to shift the server-centric access pattern to a network-centric access pattern, to realize the capacity expansion and the performance upgrade, especially the seamless data sharing among multiple servers. This naturally promotes the development of the network attached storage (NAS), by decoupling the storage and computing functionalities in any given network.

NAS connects storage devices to a standard computer network (e.g., Ethernet), which includes memory devices (e.g., hard disk arrays) and dedicated servers. There is a specific operating system (usually an optimized Unix/Linux operating system) running on a dedicated server. The dedicated server acts as a remote file server, providing network file sharing protocols such as NFS, SMB/CIFS, FTP. NAS is designed for file-level storage architecture such that other application servers in the network can remotely access the file system in storage devices through the network. After specific optimizations, the file system can support multiple file formats. Therefore, in NAS, all application servers with different operating systems can realize file sharing through the network file system. Compared with SAN, NAS provides both storage and a file system, while SAN provides only block-based storage and leaves file system concerns on the "client" side. On the other hand, NAS is a file-level storage system, thus the storage rate is relatively low compared to block-level SAN. In summary, NAS aims to offer efficient file sharing services, and is suitable for large file transmission in relatively small-scale networks.

SAN maintains high-speed and reliable interconnections between storage devices and servers, and forms a dedicated area network for data storage among the storage devices. The front-end server can access the back-end storage devices through the network. Each storage device doesn't belong to any single server anymore; instead, all storage devices share resources with all servers. According to the used protocols and communication techniques, the current SAN architectures can be categorized into two categories, including the SAN with the optical fiber communication interface and the SAN with the ordinary communication interface. SAN architecture has a lot of advantages. First, servers access the storage devices through the network, therefore multiple servers can store data to different storage devices simultaneously. In this way, storage devices are decoupled from servers; thereby, storage devices and the servers can be built and updated independently. Meanwhile, the storage devices can be shared efficiently by diverse applications, ensuring high storage utilization even if different applications exhibit unequal storage requirements. After more than 10 years of development, SAN has been quite mature and has become the de facto standard of the industry. However, for PB-level data storage needs, the SAN architectures still fall short in the scalability of the capacity and performance.

Nowadays, the consecutive development of the Internet of things (IoT) makes it possible for human beings to carry out fine-grained and continuous observation of the physical world. All kinds of sensors continuously collect various types and large volume sensing data about the physical world. These data impose serious requirements on the massive storage space, and the efficient data management and analysis techniques. The new applications, such as IoT applications, large-scale online services, enterprise fundamental services, gave birth to large-scale data centers with hundreds of thousands of or even millions of servers. Although data centers provide a tremendous amount of computing and storage resources, it is still non-trivial to handle the problems that occur in large-scale data centers, such as resource reuse, resource management, fault tolerance, etc.

To this end, Google has designed massive and performance data centers, which replace traditional dedicated devices with commodity servers, storage devices, and networking equipment. After attaching additional disks to each server, a large-scale distributed file management system, i.e., GFS [2], is implemented across the whole data center. Due to the low reliability of such commodity devices, it is inevitable for data centers to occur device failures. To improve the reliability of distributed file system, GFS implemented the data replication mechanism to achieve file system-level data redundancy. This kind of network storage architecture naturally avoids the bottleneck of SAN, and can realize the linear expansion of performance and capacity. In addition, Google also designed large-scale computing framework MapReduce [3] and large-scale distributed database BigTable [4] for its data centers. MapReduce divides computing operations into two stages, including the Map and Reduce stages. Each stage employs a portion of servers in the data center to finish a set of computation tasks. According to the data locality principle, each computing task should be loaded to the nearest node, which stores the data to be processed, so as to minimize the data transmission overhead inside the data center. BigTable is competent to process PB-level data in both a fast and efficient manner, resolving the storage and management problems of massive data.

Google uses a great number of commodity servers to build large-scale data centers and guarantees efficient collaboration, on-demand scheduling and resource sharing among each data center. The behind design and management rationales are considered as milestones of the new generation data center. Such a data center is usually composed of tens of thousands of servers, which are interconnected to form a distributed computing and storage network with a specific network topology. The new generation of data centers has attracted the great attention of Internet companies and the research community because of their inherent scalability and fault tolerance. At the same time, Amazon, Google, Microsoft, Facebook and other large-scale Internet companies have also built large-scale data centers around the world, which can accommodate tens of thousands or even hundreds of thousands of servers. With the quick development of modern data centers, many efforts have been made to improve the design technologies from different aspects. Taking the data center network as an example, the Internet Engineering Task Force (IETF) established a working group, named Software Defined Networking (SDN), which regards the data center network as a main application scenario. Besides, IEEE also started up a task group, the Data Center Bridge (DCB), for the data center network. CISCO, Juniper Network, HUAWEI and other equipment manufacturers launched various switch products for data centers.

1.1.2   Demands for Data Centers from Cloud Computing

The expose of information in modern society, the maturity of resource reuse technology and the popularization of the broadband network have jointly promoted the birth and development of cloud computing. Cloud computing platforms, such as Amazon's EC2, Google's AppEngine and Microsoft's Windows Azure, have been commercially available to offer on-demand use of hardware and software. Cloud computing has become a de facto new computing paradigm. To meet the huge market demand, traditional IT enterprises also accelerate the migration of their services to the cloud.

Cloud computing is a commercial computing model. It depends on the virtualized data center to provide convenient, flexible, and cheap online services for Internet users and enterprise employees. This fact clarifies the essential status and role of the data center as the core infrastructure of cloud computing. Gartner estimated in a July 2016 report that Google at the time had 2.5 million servers across multiple data centers around the world. In April 2011, Facebook unveiled its data center in Oregon with tens of thousands of servers. This data center utilized some energy efficiency and emission reduction technologies. At present, far-sighted local governments in China are building data centers to promote the cloud computing industry, such as Guiyang data center, Hohhot industrial base for cloud computing, and Wuxi cloud valley. All telecom operators in China have launched their cloud computing strategies with their Internet data centers. Internet companies such as Baidu, Alibaba and Tencent also developed large-scale cloud data centers to provide better cloud computing services.

The cloud platform is the foundation of cloud computing services. It manages a huge amount of physical resources inside a data center and is responsible for the global allocation and scheduling of these resources. Thus, it is possible that various cloud services can compete for resources while running steadily in the complex cloud environment. Most cloud platform uses many virtualization technologies to integrate various resources, to provide the Infrastructure as a Service (IaaS), Platform as a Service (PaaS), and Software as a Service (SaaS) to users in a transparent way.

IaaS aims to provide users with virtualized computing resources, storage resources and network resources inside data centers, and dynamically adjusts the assigned resources to each user on demand. For example, Amazon's EC2 [6] provides rental services of virtual machines for small and medium enterprises to flexibly build-remote IT infrastructures based on the virtual machine. EC2 is a representative application model of IaaS, and is currently one of the most mature IaaS systems in production. We also note that some open source cloud computing projects, e.g., Nimbus [7], Eucalyptus [8], and Opennebula [9], provide similar services as EC2.

PaaS provides a platform allowing users to develop, run, and manage applications on the basis of a series of fundamental services, such as databases, data processing and software development environments. In this way, users can customize their applications on the PaaS platform, without the complexity of building and maintaining a local development platform. Google App Engine (GAE) [10] is a representative PaaS platform. It offers users the development and operation environment based on Java and Python. In addition, Google also releases some development tools to help programmers to develop Web programs. Windows Azure Platform (WAP) [11] is similar to GAE. It deploys the virtual machine as a running environment and provides data storage services to facilitate users to develop and implement applications.

SaaS tries to provide end-users with customized software services so that users can access them in the cloud without local installation. Compared to IaaS and PaaS, SaaS is more suitable for ordinary users. Many SaaS services are established based on IaaS and PaaS platforms. Google Apps and Salesforce are two widely used SaaS platforms. Google Apps provides users with web applications with a similar function of desktop software. Users no longer need to download software, and just access them through the Internet. Unlike the popular free Google SaaS platform, Salesforce aims to deliver a commerce platform for customized applications.

Overall, large-scale scalable hardware resources are the basis of resilient cloud computing services. The appearance of a data center aims at providing a large number of physical resources and then deploying dedicated cloud platforms to offer users multiple types of cloud services. The public cloud data centers usually provide the rental services of IaaS and PaaS first, and further offer the basic environment for the SaaS development and deployment. Private cloud data centers run specific network services for dedicated companies or departments only; while hybrid cloud data centers deliver more flexible and diverse services. For example, Google's cloud data center not only offers front-end search services for customers, but also runs back-end analysis and mining tasks of massive webpage data. The emergence of the cloud computing model significantly extend the usage of data centers. In turn, the development of data centers also build a solid foundation for the polarization of cloud computing.

1.1.3 Demands for Data Centers from Big Data Applications

In March 2012, the Obama administration announced an $200 million investment plan to launch the Big Data Research and Development Program. This program is another key science and technology strategy after the Information Superhighway Program launched in 1993. This will undoubtedly have a far-reaching impact on the development of science, technology, and economics in the future. On the other hand, China, the European Union, Canada, the Republic of Korea, Singapore and other countries or regions have launched similar big data development strategies. The definition of big data is not unified by academia. One representative definition is that big data is a collection of data, which cannot be captured, managed, and processed using conventional software within a bearable time period. International Data Corporation (IDC) defines big data with four features, namely, massive data scale (Volume), diverse data types (Variety), huge data value (Value), and fast data flow (Velocity).

The data from Internet of Things (IoT) applications represents a major source for big data. IoT aims to expand the existing Internet and telecommunication infrastructures to interconnect ordinary devices, and facilitate communication between the devices themselves and/or between the devices and humans. The IoT architecture consists of four layers, namely, the perception recognition layer, the network construction layer, the management service layer, and the application layer [12]. In the perceptual recognition layer, a large number of sensing units continuously generate time-series observation data, hence act as important sources of big data. When the large volume of sensing data is transferred to some management services via the network layer, it is necessary to store, analyze and process these data. As the quick growth of IoT applications, the data generated by all sensors leads to an exponential growth in the storage and computing requirements. The data center in the management service layer acts as a powerful resource platform for IoT applications. It offers storage services as well as processing services for IoT applications. In 2015, IDC predicted that IoT would require $7.5 \times$ cloud resource capacity in the next four years.

The birth of big data is a natural result of the development of information technologies. The bulk of big data generated comes from various sources, such as IoT, Internet, social network, social public domain and beyond. Big data has become the third-largest social resource besides matter and energy. The key challenge is how to quickly extract value from massive and heterogeneous data. Big data involves many industries and various domains across the globe, and faces the following common problems: (1) the collection and preprocessing methods of big data; (2) the storage and management methods of big data; (3) the analysis methods of big data; (4) the architecture and platform of the big data system. As an information infrastructure, the data center offers a foundational platform for various big data applications and has the natural advantages of addressing the above challenging issues.

1.1.4 The Development of New Generation Data Centers

Traditional data centers incur a hand of disadvantages, such as the low resource utilization, the long delivery cycle time of new services, etc. Such issues make them fall short of supporting the applications in the fields of cloud computing, mobile Internet, Internet of things, and big data processing. The emerging new technologies and applications impose serious requirements on the next-generation data centers, such as the large-scale cloud data centers, and geo-distributed data centers. In addition, the business model of the data center has been shifted from traditional resource leasing to the cloud service renting. This new business mode exhibits new features, including high elasticity, high availability and the high performance, etc. Currently, many IT companies, telecommunications operators, and government departments have provided many cloud data center services. In the future, data centers will keep embracing virtualization, modular design, energy-saving, and automatic maintenance technologies.

1. Virtualized data centers (VDC). Virtualization has the advantages of flexible allocation, quick updates, and high resource utilization. As reported in literature, the resource utilization of data centers is only 20–30% on average, while the power consumption in most idle cases still reaches about 60% of the peak value. The appearance of VDC accelerates the development of cloud computing. It provides a high level abstract to virtualize and integrate the physical resources and thereafter offers a unified interface for upper-layer applications, such that various resources inside a data center can be dynamically allocated. As a result, VDC naturally guarantees high resource utilization, less energy consumption, and low operation cost. Virtualization technologies, including the traditional computation and storage virtualization, as well as the emerging network and IO virtualization, play an increasingly important role in cloud data centers.

2. Modular data centers (MDC). The modularization design methodology brings remarkable manageability and operation efficiency to data centers. It can satisfy the requirements of building customized data centers in a flexible and expandable way, hence support the increasing amount of data center applications. Specifically, MDC involves many challenges in practice, including customized data center modules, the standard assembly process of modules, efficient delivery strategy, and on-demand module expansion and replacement, et al. MDC naturally eases the construction, maintenance, deployment and expansion of data centers. If the price of an MDC could decline further in the future, it will be one of the most important choices for small and medium-sized enterprises.

3. Green and energy-saving data centers. Energy-saving is a major development trend of future cloud data centers. It aims at introducing advanced technologies to improve energy efficiency, thus saving total energy consumption. Energy-saving strategies are usually utilized across the whole room, cooling system, and IT facilities. Consequently, a data center can maintain a stable circumstance for involved devices with the ambition of energy reduction. Available energy-saving strategies include water cooling refrigeration technology, high-density power distribution technology, high voltage power supply technology in high-density computer cab-

inets, airflow control technology, intelligent environment monitoring technology, etc. In addition, Such energy-saving strategies should be adopted across the planning, design, deployment, operation, maintenance, and management phases of the cloud data center.

4. Data centers under automatic operation and maintenance. With the quick development in recent years, data centers accommodate various applications, thereby the operation and maintenance of data centers become extremely heavy and complex. In large data centers, traditional operation and maintenance methods are unable to resolve the low efficiency, high labor cost and error prone challenges. The emerging AIOps method aims to introduce a complete and automatic management and control scheme for data centers in the process of configuration management, performance monitoring, et al. This will significantly improve the automatic level of production data centers, and significantly reduce those daily manual operation and maintenance work. Furthermore, this would standardize involved operation and maintenance activities, and consequently avoid human errors.

As the core information infrastructure, the next-generation data center will play a more important role in the future information and intelligent Era. According to the 2018 report of the international data company (IDC), the sum of the world's data will grow from 33 zettabytes in 2018 to 175 ZB by 2025, and will be accommodated by public large-scale data centers. The data center also motivates the quick development of novel computing and storage technologies. Many IT companies and institutions have built many geo-distributed data centers, interconnected by dedicated WAN, to offer various services for customers.

1. Google's Data Centers
   By 2020, Google has built many large-scale data centers around the world with collaborations from other companies. At least 13 significant google data center installations are located in the United States, 5 in Europe, 1 in South America and 2 in Asia. These data centers support many business needs from search, YouTube, e-mail, et al. Google has released its own cloud service platform, i.e., Google Cloud Platform, which provides users with four types of products: computing, storage, big data and services. Besides, Google has launched at least 11 types of services to cover all aspects of social life, including multimedia, mobile applications, web applications, e-commerce, software development, big data, financial services, games, Internet of things, genetics and security. Based on its advanced data centers, Google can support at least one billion query results within several milliseconds, maintain 6 billion hours of YouTube video playback per month, and provide storage service for more than 1 billion Gmail users [13].

2. Microsoft's Data Centers
   By 2020, the global cloud infrastructure of Microsoft is comprised of 160+ physical datacenters, arranged into regions, and linked by one of the largest interconnected networks in the world. Such data centers offer high availability, scalability, the latest advancements in cloud infrastructure. With such data centers, the Microsoft Azure cloud platform provides software as a service (SaaS), platform as

a service (PaaS) and infrastructure as a service (IaaS) to the global users. Microsoft lists over 600 Azure services, which support many different programming languages, tools, and frameworks. Azure has attracted many giant users, including some famous organizations and companies, to build, manage, and deploy applications using their favorite tools and frameworks, so as to meet their business challenges.

3. Amazon's Data Centers

   Amazon is currently the largest provider of cloud computing services in the world, accounting for about 34% of the market. However, Amazon did not release any report about the number and locations of its data centers. WikiLeaks published a list of AWS data center locations in 2015 [1]. In the US, Amazon operates in some 38 facilities in Northern Virginia, eight in San Francisco, another eight in Seattle and seven in northeastern Oregon. In Europe, it has seven data center facilities in Dublin, Ireland, four in Germany, and three in Luxembourg. There are 12 facilities in Japan, nine in China, six in Singapore, and eight in Australia. It also hosts infrastructures in six sites in Brazil. In terms of IaaS services, Amazon AWS provides a wide range of core IaaS services, including computing, storage and content delivery, databases, the Internet of Things, et al. Moreover, AWS provides abundant PaaS cloud services such as data analysis, enterprise-level applications, mobile services and the Internet of things. AWS adopts strict standardized designs to serve all the workload of near one million users across more than 190 countries and regions. Amazon AWS [14] has attracted many well-known companies and organizations around the world, including Reddit, Netflix, Coursera, Siemens, Adobe, 360, Oppo and more.

4. Facebook's Data Centers

   Facebook is the biggest social network worldwide since it has over 2.6 billion monthly active users as of the first quarter of 2020. Additionally, Facebook is the sixth-busiest site on the internet in July 2020 according to Alexa. To support this already massive and still growing user base, Facebook has built numerous gigantic data centers around the world. Facebook has not stopped building new data centers and continue scaling the capacity of existing data centers, so as to make its data center infrastructure even more powerful and efficient. Facebook has expanded to a total of 15 data center locations, with new centers announced. Unlike other companies, Facebook makes its server, data center design open-source to the public. Facebook's data centers mainly support its social networking services. In fact, Facebook keeps collaboration with other cloud service providers to handle part of its social network traffics.

5. Alibaba's Data Centers

   Alibaba Cloud, also known as Aliyun, is a Chinese cloud computing company, which provides cloud computing services to online businesses and Alibaba's own e-commerce ecosystem. Alibaba began to build data centers to support Ali cloud services around the world since 2014. In 2014, Alibaba built five data centers in Hangzhou, Qingdao, Beijing, Shenzhen and Hong Kong to occupy the market of cloud services in China. In 2015, Alibaba built its sixth and seventh data centers in Silicon Valley and Dubai. By 2020, Alibaba operates in 22 data center regions and more than 63 availability zones around the world. Each data center region consists of multiple available zones, while each available zone contains at least one data center. Alibaba provides secure and stable IaaS, PaaS, SaaS for various applications, such as e-commerce, big data, database, IoT, object storage, and data customization. Besides the cloud data centers, Alibaba also established its cloud content delivery network. This network has connected 2300+ nodes in China and 500+ nodes in more than 70 countries, and offers 130Tbps bandwidth.

6. Baidu's Data Centers
   Besides the traditional Internet search business, Baidu proactively expands and promotes cloud services. At present, Baidu operates data centers covering more than 10 regions in China, including Beijing, Baoding, Suzhou, Nanjing, Guangzhou, Yangquan, Xian, Wuhan and Hong Kong. It is reported that the data center in Yangquan hosts over 150,000 active servers. This data center uses a more environmental-friendly photovoltaic power generation technology to provide green and clean energy. Consequently, the annual PUE of Baidu Yangquan data center was 1.1 in 2017 and it decreased to 1.09 in 2018. Baidu's Nanjing data center employs a series of advanced technologies. For example, It utilized ARM-based servers with customized architecture and its own SSD products. Baidu's cloud products include cloud servers, object storage, content distribution network, MapReduce, machine learning, audio and video transcoding, and beyond.

7. Tencent's Data Centers
   By 2016, Tencent has operated several large-scale data centers distributed in Tianjin, Shenzhen, Chongqing, Shanghai, Guangzhou, Hong Kong, and North America, to support its cloud services. Among them, the data center in Tianjin has the largest scale with 200,000 servers. It is the largest data center in China and even in Asia. The data center is partially composed of a series of high-density, energy-efficient, easy-deployable micro modules. Each micromodule is equivalent to a traditional data center, which includes a refrigeration system, a power supply system, as well as a network, monitoring and other independent units. By 2020, Tencent cloud infrastructures are built based on 54 availability zones across 27 regions globally. Every region is an independent geographical area, which ensures maximum stability and fault tolerance among different regions. Within each region, there are multiple locations isolated from one another, referred to as availability zones, which are physical data centers. Tencent plans to gradually increase the number of regions to have greater node coverage.

8. Huawei's Data Centers
   By 2020, Huawei cloud infrastructure operates 6 regions in China and other 6 international regions. Additionally, Huawei released the smart data center solution, which provides a modern foundation for distributed cloud applications. This solution helps customers build agile, reliable, and energy-efficient data centers by integrating smart, plug-and-play micro modules on demand. Recently, Huawei released its intelligent data center service solution. This service can help customers build and operate the world's high-reliability, green and intelligent data centers. It aims to reduce the PUE by 8.

1.2 Fundamental Services Offered by Data Centers

As a fundamental information infrastructure, the data center usually offers basic storage and computation services for all users. Hence, it has been widely used by many application fields, such as big data, Internet of things, scientific computing, artificial intelligence, hence bringing huge commercial and social benefits.

1.2.1 Storage Service

Large-scale data centers prefer to provide scalable and reliable online storage services, which seamlessly satisfy the storage requirements of various Internet applications.

Google file system (GFS) is the core part of Google's networked storage system. Many critical applications rely on the storage services provided by GFS, such as Gmail, Picasa and Google Doc. The initial motivation of GFS is to resolve the common challenges of massive data storage and processing faced by search engines. Therefore, in addition to the general functions of distributed file systems, GFS has achieved the following improvements: (1) GFS usually needs to handle large files, hence it uses 64 MB as its basic storage unit; (2) GFS updates files via the append operation, therefore, it is very important to optimize the performance of the append operation; (3) the volume of metadata of files could be significantly reduced after setting the storage unit as 64 MB.

GFS presents the above special design according to the requirements of its own applications. However, for other applications, it is necessary to adjust the GFS framework based on dedicated characteristics and requirements. Additionally, node failure is very common for any data center; hence, GFS employs strong fault tolerance mechanisms as follows. The basic idea is that at least three replicas of each block are stored across servers. In this way, the block write operation may lead to the data inconsistency problem across multiple replicas. To tackle this critical issue, GFS treats all replicas of each block as a chain. When a new block is generated, GFS would write all involved replicas along with the chain and treat the entire write process as a transaction. When an existing block needs to be updated, the involved replicas would be processed in a chain manner, and the version number would be overwritten only if the entire chain has been successfully updated. On the basis of the GFS, a number of important open source storage projects have emerged, including Hadoop File System (HDFS) [15], KFS [16], Sector/Sphere [17] and so on. Meanwhile, more and more commercial storage systems utilize HDFS as their basic building block.

Amazon offers another typical networked storage system, Simple Storage Service (S3), which provides users with a highly reliable and private data storage environment. The infrastructure of S3 is a scalable system, consisting of a large number of commodity storage devices. This design philosophy is exactly the same as the GFS. The SkyDrive of Microsoft offers free online storage, while the Upline of HP charges its users with the data backup service. The Atoms of EMC is a cloud storage system based on object-oriented storage techniques. It also relies on commodity storage devices and provides a set of services, such as data compression, data deduplication, and disk hibernation. In addition, more and more cloud storage services appear in China. For example, Alibaba and Baidu started to offer free and paid networked storage services.

1.2.2 Computing Services

At present, many IT companies have built their own computing platforms based on data centers. The mainstream solutions rely on distributed computing frameworks, such as MapReduce [3] and Dryad [18]. The MapReduce offered by Google is the most representative one. MapReduce is a reliable, efficient and scalable networked computing model based on GFS. It usually includes three phases: Map, Shuffle, and Reduce. The Map and Reduce phases execute the large-scale computing tasks with massive computing nodes in data centers. Each computing node is responsible for serving a set of data processing tasks in parallel. The task assignment strategy in the Map phase needs to consider the data locality issue, i.e., assigning each computing task to that node storing the involved input data. As a consequence, unnecessary data transfer across nodes can be avoided. Google relies on the MapReduce service to perform large-scale data processing tasks in daily work. This joint model with both networked storage and networked computing is becoming more and more popular, and have been widely used in many scenarios.

Hadoop is a most famous open source project, which provides similar functions as the GFS and MapReduce. Nowadays, Hadoop has been widely applied in many fields and has been commercialized by many enterprises. After the emergence of Hadoop, researchers have developed many large-scale data analysis systems. In 2010, Google released the technical details of its Caffeine [19], Pregel [20] and Dremel [21], explaining how its data center can support massive network applications. Caffeine is dedicated to supporting the web search engine of Google such that Google can add new URL links to its index system of web sites more quickly. Specifically, Caffeine store a huge volume of URL indexes in a distributed database, called BigTable. Pregel is a large-scale distributed graph computing framework, which is especially employed to handle the computing problems in the web link analysis, social data mining and other applications. Pregel adopts an iterative computation model. In each iteration, each vertex processes the messages received from the previous iteration, and then sends the resulted message to other vertices, and updates its own state and the network topology. Dremel is an interactive data analysis system, which can well compensate for the MapReduce computing frameworks. Dremel is able to run across thousands of servers, allowing multiple query schema over large-scale data at an extremely fast speed. According to a report by Google, Dremel can complete the query request over 1PB data in about 3 s, while MapReduce takes longer time than Dremel to respond to the same query.

The two-phase computing model of MapReduce simplifies the programming interface for users, eases the programming procedure, optimizes the resource scheduling and supports fault-tolerant data processing. However, its one-fold programming model limits its flexibility to support more general applications. To this end, Microsoft has developed another efficient distributed computing framework, named Dryad [18], which provides a new solution besides MapReduce. Dryad shares the same design philosophy with MapReduce. The difference between them is that, MapReduce divides the user logic into two phases, i.e., Map and Reduce; while Dryad only has one single abstraction, i.e., Vertex. Users realize their computation logics by implementing customized Vertex nodes, and nodes exchange data via various of data channels. For achieving generality, Dryad neither distinguishes the operation phase in its computation model, nor defines the format of exchanged data among computing nodes. Instead, Dryad leave each pair of nodes to tackle the format-compliant issue on demand. This would increase the difficulty of programming to some extent, but brings significant flexibility of programming. DryadLINQ [22] is a distributed computing language, which could transform programs written by LINQ into programs of Dryad. Thus, programmers can easily implement large-scale distributed computing over thousands of computers, and would not care about the details about the distributed computing systems.

1.2.3 Big Data Applications

There are many kinds of big data applications accommodated by data centers. As aforementioned, many specific distributed computing frameworks have been designed for diverse big data applications. Among them, Hadoop is one of the most representative distributed computing systems. The MapReduce in Hadoop is suitable to support Internet applications, such as ranking among web pages and efficient search in complex networks. Spark [23] is an open-source, in-memory processing systems used for big data workloads. It can achieves high performance for both batch and streaming data, using a state-of-the-art DAG scheduler, a query optimizer, and a physical execution engine. GraphLab [24] is a representative graph computing framework. It is suitable for solving big data applications, whose tasks can be formalized as a graph computing model and require iterative calculations. Storm [25] is an open-source real-time computing framework (also known as a stream computing framework) with the benefits of low latency, scalability, and fault tolerance. It can well satisfy the requirements of latency-sensitive applications, such as on line anomaly detection, online machine learning and so on. Parameter Server is a distributed machine learning framework. It specializes in solving machine learning applications, where the tasks require a large number of iterative computations and involved nodes exhibit the data dependency.

Hadoop is a distributed system infrastructure, released by the Apache foundation. It mainly consists of two core parts, i.e., the distributed file system HDFS and the distributed computing framework MapReduce. MapReduce has great ability of fault tolerance and significantly simplifies the workload of programmers. In the developing process, programmers only need to implement their Map and Reduce functions. Other issues involved in the distributed system are resolved by the MapReduce framework, such as the communication mechanism and the fault tolerance mechanism. This programming model is especially suitable for the web search and sorting application, and other search applications with the input of large-scale data sets. Hadoop is an open-source system, and is continuously improved by many volunteers; hence, more and more components have been released. As a general distributed computing framework, MapReduce has been widely utilized in many fields.

Spark is another general distributed computing framework, released by the Berkeley University. Unlike MapReduce, Spark is a memory-based distributed computing framework. That is, the intermediate output and final results can be stored in memory, thereby it is not necessary to read (write) data from (to) HDFS. As a result, Spark guarantees better performance for iterative data processing. Specifically, Spark uses an resilient distributed dataset (RDD) as its basic data structure. The all tasks in each iteration only manipulate the same RDD, thereby Spark exhibits better performance than MapReduce for iterative computations. Therefore, Spark is very suitable for data mining and machine learning applications, which require more iterations of computation and access large volume of data. By contrast, the benefit of Spark is relatively limited for computation intensive applications with small amount of input data.

GraphLab is an open-source graph computing framework, launched by Carnegie Mellon University. GraphLab aims to model and implement iterative algorithms for machine learning applications. Moreover, GraphLab tries to guarantee the data consistency and the computing performance. Although GraphLab was initially developed to handle large-scale machine learning tasks, it is also applicable for many data mining tasks. GraphLab can be deployed on many different computation environments, include a standalone system with multiple processors, a cluster or a data center.

Many data mining and machine learning algorithms exhibit two common features. First, the involved computing nodes maintain a strong data dependence. Usually, a large machine learning model contains billions of parameters, and the computing nodes need to communicate and share the learned parameters with each other. Second, a machine learning job usually needs to perform multiple iterations, therefore it is difficult to realize the computation parallelism. The distributed computing frameworks like Hadoop and Spark can efficiently execute tasks without data dependence, but remains inapplicable for tasks with high data dependence. Compared with Hadoop and Spark, GraphLab aims to support general machine learning algorithms, and can achieve one or two orders of magnitude speedup.

Storm is an open source framework for distributed stream data processing. It is especially suitable for time-sensitive applications with massive data. With the continuous growth of data scale, real-time data processing has become a major challenge for many applications. These applications include the anomaly detection in multimedia, real-time recommendation in e-commerce, user behavior analysis in search engines, online machine learning, etc. The Storm framework consists of a master node and a large number of worker nodes. The master node executes the background program

Nimbus to manage all worker nodes and assigns computation tasks. Each worker node runs the program Supervisor for receiving and executing tasks. Additionally, the component Zookeeper coordinates the Nimbus and Supervisor. In Storm, data are inputted into the component Spout, and then the Bolt component implements the real-time processing task.

Parameter Server is a representative distributed machine learning framework. For machine learning algorithms, it is very common that a large number of computing nodes share the learned parameters. Besides, all involved nodes are required to conduct multiple iterations to complete the learning task. It is well known that Hadoop needs to store the intermediate parameters to their distributed file systems. As a consequence, the frequent read and write operations will surely slow down the learning speed. Spark is an in-memory computing framework, and adopts a data consistency protocol based on its block synchronization mechanism. Therefore, a computing node has to wait for the completion of tasks on other nodes before the next iteration. Therefore, it is difficult for Spark to quickly complete a machine learning task with large-scale parameters. Many graph computing frameworks, e.g., GraphLab, can quickly deal with some machine learning tasks; yet it can only handle tasks that can be converted into graph computing models. For other tasks that cannot be represented as graph calculation models, all existing graph computing framework are still helpless. Parameter Server is a dedicated framework for general machine learning tasks. The server node collects the learned parameters of all worker nodes in each iteration, and send back the aggregated results of all parameters to each worker node for the next iteration. In each way, all worker nodes actually share all learned parameters with each other at the end of each iteration. With this design, the Parameter Server framework can significantly speed the completion of the machine learning tasks.

With the emergence and development of various distributed computing frameworks, it has become a research hotspot to process large-scale scientific data using distributed computing frameworks. Based on Hadoop, researchers have designed Sci-Hadoop [26] and Hadoop-GIS [27] systems for scientific computing, using large, heterogeneous and multi-dimensional datasets as the input. These systems transform the file-oriented storage model of Hadoop as a novel storage model for multi-dimensional data so as to enable efficient Spatio-temporal data queries. They provide users with domain-specific data manipulation interfaces. Accordingly, users can focus on their applications without worrying about the details of the underlying distributed storage and parallel computing technologies.

1.3 Challenges for Data Center Networks

As aforementioned, data centers have become the fundamental infrastructure for countries as well as enterprises. As the basic component, data center network (DCN) plays an essential role for the design of data centers. DCN aims to connect all switches and servers together using an outstanding topology, so as to achieve better network performance. In this way, data center can achieve comprehensive advantages in terms of the computation, storage and communication. That is, DCN is not only a bridge to interconnect large-scale servers, but also a basis for offering storage and computation services. Many applications hosted by data centers often cause intensive data interaction across multiple servers. DCN becomes the bottleneck of data centers, and affects user's experience directly. Therefore, it is urgent to bring more innovations into the field of DCN, so as to speed the development of data centers and its applications.

On the other hand, the data center network has become an important part of the Internet. According to a report from CISCO, the traffic of data centers is mainly composed of three parts. About 76% of the traffic is caused by the interactions among servers inside the data centers. Around 17% of the traffic is caused by the interactions between wide-area users and remote data centers. The rest 7% of the traffic is caused by the interactions among data centers. As CISCO stated that, by 2021, the data center network traffic reached 20.6 ZB (1 ZB= 10^{12} GB). In other words, the traffic faced by data center networks becomes the majority of Internet traffic.

Compared with WAN, DCN has distinct features, such as the centralized management and control, which can ease the exploration and deployment of emerging network technologies in DCN. One of the main obstacles for network innovations is the coordination and gaming among the Internet service providers (ISPs). As a result, the ISPs are not willing to deploy new technologies without strong incentive mechanisms. By contrast, a data center is usually owned and managed by a single cloud service provider. For achieving the better performance and venue, the data center operators customize their network architectures and protocols and implement advanced network technologies, according to users' demands. For example, Google widely deploys the SD-WAN across data centers for flexible network functions.

The research about DCN has become a highlight field in academia as well as the industry communities since 2008. There still exist a series of fundamental challenges for speeding the development of the data centers. Specifically, this book focuses on two challenges, including the topology design and the collaborative management of traffic inside each data center.

1.3.1 Customization of Network Functionality

The early data center networks can only provide limited functions and network services. The fundamental reason is that the control plane and data plane are tightly coupled, therefore only standard or predefined packets can be forwarded by network devices without any flexibility. If users need to customize network functions and services, which are not supported by the current DCN, they must purchase and redeploy dedicated network devices. This upgrade strategy is surely time-consuming and costs a large amount of investment. Moreover, replacing network devices has to interrupt the flow transmissions and all involved network services. Nowadays, the emerging network applications attach a variety of requirements to data centers. To improve the quality of services, the data center providers have to configure network functions dynamically according to the application requirements. Additionally, the traffics inside data centers are usually unpredictable, and the failures of commodity network devices are very common. These features put forward new requirements to the dynamic and flexible deployment of network functions. Therefore, it is challenging to realize network function customization in traditional DCNs.

Software-Defined Networking (SDN) is a promising network technology emerging in recent years [28]. SDN has the following two core ideas: (1) enhancing the programmability of hardware, which enables fast configuration of new network functions to satisfy the varying requirements; (2) decoupling the control plane from the data plane of the entire network and collecting the control functions at the controllers, which significantly eases the network management and control. It is still an open question whether the SDN technology can be applied to the Internet or not, yet DCN is a perfect environment for the deployment of SDN technologies. A software-defined DCN is an infrastructure-level guarantee for the upgrade of network performance, and the network sharing among multi-tenants. The OpenFlow protocol proposed by Stanford University is a representative SDN protocol. However, due to the complex data forwarding and limited functionalities of the data plane, researchers are exploring other alternative SDN architectures.

The programmable networks and software-defined networks have offered the necessary conditions for data centers to realize customized network functions. However, the scarcity of computing, memory and other resources of traditional network devices severely limits the types of customizable network functions. The reason is that data center attaches increasingly complicated and sophisticated operations onto each flow, e.g., data stream caching, data stream analysis, and intra-stream processing. These network services would occupy a large amount of hardware resources. For example, deep analysis of data flow consumes a vast amount of computing and memory resources, and customized flow routing requires frequent queries upon routing tables, thus occupying memory resources. The contradiction between the network function customization and scarce hardware resources of network devices has brought unprecedented difficulties and challenges to DCN.

1.3.2 High Scalability of Data Center Networks

Datacenter tries to satisfy different types of access requirements for computing, storage, communication and other resources. To this end, data centers must be expanded on demand, in terms of computing, storage, and network resources. That is, the data center has to efficiently interconnect tens of thousands or even more servers and network devices using dedicated network topology. Traditionally, we improve the performance of a single switch to connect more servers, by increasing the number of switch ports and hardware line rate with terrible hardware investment. These kinds of methods are unable to meet the scaling requirements of DCNs. Therefore, it is urgent to replace the traditional scaling methods with a variety of high scalable network topologies, which interconnect more switches and servers to achieve on-demand expansion of the computing performance and storage capacity.

Data centers, which connect a large number of servers through specific network topologies, are the infrastructure for various applications and services. After a huge amount of data is partitioned and stored across servers, many data-intensive jobs would be assigned to multiple servers to perform involved tasks concurrently. For each task, it is necessary to acquire and process a large volume of distributed data. Therefore, a great number of east-west traffics are triggered inside a data center, which has replaced the traditional north-south traffics as the dominated traffics. This attaches severe challenges to the aggregate bandwidth of DCN. All existing DCNs can ensure that the computation and storage capacity increase linearly with the number of servers. However, they differ in the network performance, such as the aggregate bandwidth and transmission delay, which directly affects the availability of a data center. Therefore, DCN is not only a bridge to interconnect massive servers, but also a key component to improve the performance of cloud services.

Traditional data centers usually adopt tree-like network topologies, each of which is composed of three layers of switches, i.e., edge switches, aggregation switches and core switches. However, it has been proved that this kind of topologies cannot satisfy the increasing requirements of the cloud services. Therefore, it is urgent to design novel and scalable DCNs. Although many scalable DCNs have been proposed recently, they are built on top of homogeneous devices, lacking considerations about failures of devices and links. Consider that the scale of a production data center is incrementally expanded along with the increase of applications and users. It is very common that heterogeneous servers and network devices would be inevitably introduced into the DCN network during the expansion and maintenance process of a data center. To improve the resource utilization, the data center network must be capable of interconnected heterogeneous devices in a scalable way. In addition, for less hardware investment and high performance-price ratio, data centers prefer to use commodity equipment to replace expensive dedicated ones. While commodity equipment could significantly reduce the investment and ease the operation, they exhibit lower availability and reliability than the dedicated devices. Therefore, how to build reliable data centers, using a large number of heterogeneous yet commodity servers and network devices, is an essential challenge for the design of a scalable DCN.

1.3.3 Efficient Multiplexing of Network Resources

One of the core values of the data center is its efficient statistical multiplexing of resources. For a given data center network, it is necessary to optimize its network utilization via software technologies, so as to improve the performance of applications and guarantee the user experience. The network protocols adopted by data centers are originally designed for the WAN environment. Unfortunately, the network environment of data center differentiates from that of WAN, such as the burst and unpredictable traffics, extremely high end-to-end bandwidth, very low end-to-end delay, etc. As a consequence, the network utilization inside a data center is remarkably low under the support of those network protocols. For example, traditional OSPF cannot fully exploit the link resources in DCNs, and incur slow convergence speed. Thereby, they are not competent to meet the high-speed transmission requirements of various applications. Besides, TCP, a data transmission protocol successfully used in WAN, is also very inefficient in DCNs with high link bandwidth. The TCP Incast problem can make the bandwidth utilization of the network less than 10%, which is called the throughput collapse in data centers.

Under the software-defined customizable network architecture, it is critical to design new routing and transmission protocols to improve the resource utilization of DCNs and thus enhance the performance of upper-level applications. In addition, the software-defined network for a data center provides an opportunity for the joint optimization of network, computing, and storage resources. Note that the overall performance of a network system often relies on the dependency among its subsystems. The software-defined network separates the network control functions from the forwarding devices. Such a design is aware of the requirements of applications and data storage status through the information interactions among the controllers. Moreover, it realizes cooperative control across different types of resources; hence, improving the resource utilization and service quality. Therefore, it is necessary to investigate how to jointly optimize the network, computing and storage resources through software-defined network technologies, so as to improve the resource utilization of the data centers significantly.

1.3.4 Network Virtualization of Data Centers

For a data center, it is necessary to realize effective network sharing and security isolation since a large number of users compete the network resources. Virtualization is an important enable technology to realize the resource multiplexing and improve the security of the entire data center. Computation virtualization and storage virtualization are relatively mature enough for managing a data center. Via computation virtualization, users can access the computing resources in a multiplexing manner, without any worry about the management, maintenance and upgrade of the physical computers in a data center. In addition to the storage multiplexing, the storage virtualization technologies further guarantee the backup and security of user data. With such virtualization technologies, the concepts of use-on-demand and pay-on-demand are becoming realities and become the major service model of data centers.

In practice, many users often need more than one virtual machine to offer the required computing resources. All virtual machines allocated to each user need efficient interconnection as well as interactions; hence, they are connected via a virtual network and form a virtual data center. All virtual networks for many users actually coexist and overlay on top of the common physical data center network. The sharing characteristic of network links make those virtual networks compete for the actual physical bandwidth. For the security issue, all virtual data centers for different users should be isolated, and any pair of virtual machines that belong to different virtual data centers could not communicate with each other by default. The development of network virtualization is relatively slower than the computing virtualization and storage virtualization. At present, multiple users share the network resource of a data center in the best-effort manner. Consequently, the network bandwidth obtained by each user is unpredictable, and traffic leakage across virtual data centers is a potential and critical problem. These shortcomings would significantly damage the user experience. In summary, existing data center networks fail to provide elegant traffic isolation and bandwidth guarantees for virtual data center networks.

In the software-defined DCN, the network controller maintains the global information about the physical network and resource consumption, thus providing a flexible control platform for the management of various virtual data center networks. However, there still exist many challenging issues remained. For instance, how to deal with the setting issue if diverse users configure common IP address or MAC address, how to ensure that all traffics of a user will not leak to other users, how to realize efficient migration of virtual machines inside a virtual data center network, and how to ensure that each user gets fair bandwidth allocation in the shared network.

1.3.5   Cooperative Transmission of Correlated Traffic

Although the proposals of scalable data center networks for data centers can realize large-scale expansion and increase the network capacity, sharing network resources among multiple users still attaches severe challenges. Specifically, many big data services hosted by a data center need to frequently acquire and process distributed data. These intensive data interactions generate a large number of east-west traffics, thus the shared bandwidth becomes the bottleneck of services. At the same time, data centers usually deploy a variety of distributed computing frameworks to analyze and process massive datasets. Many of such computing frameworks utilize the stream computing models, incurring intensive data transfers of intermediate results among servers. Such inter-stage transfers would consume a lot of network bandwidth and directly increase the overall completion time of jobs and applications. For example, when Google provides a famous search service to users, it needs to perform large-scale real-time analysis and mining of massive webpage data at remote data centers. The multicast, incast, and shuffle are the most important traffic patterns since they are the majority of the east-west traffic in data centers. Besides, all involved flows of an incast or shuffle transfer are inherently correlated with each other.

Although the scalable topologies can continuously improve the network capacity of data centers, it is still important to efficiently exploit the available bandwidth of a production data center. As aforementioned, data flows in an incest or shuffle transfer are usually highly correlated, and are traditionally aggregated at specific receivers. If such data flows can be early aggregated during the transmission process, this can significantly save the bandwidth resource. However, the traditional traffic management model fails to enable this in-network aggregation requirement. Fortunately, after introducing the software-defined network technologies into the design of data centers, the cross-layer joint design can significantly reduce the transmission overhead of the correlated flows, without attaching any damage to the application performance. As a result, it is feasible for those correlated flow transfers to consume much less bandwidth. However, this design still involves the following challenges.

1. *In-network aggregation of correlated flows* In many distributed computing frameworks like MapReduce, data flows in an incest or shuffle transfer are highly correlated. Therefore, when correlated flows arrive at corresponding receivers, many aggregation operations can be executed across flows. Although these aggregation operations can significantly reduce the amount of data transmission, they are not performed during the transmission process. With the idea of in-network aggregation, the correlated data flows of incast and shuffle can be aggregated in the transmission process as early as possible, instead of aggregating them at the receivers. In this way, the consumed network resource of these correlated flows can be significantly decreased.

2. *Cooperative transmission of correlated flows* The in-network aggregation for many correlated flows, which constitutes the same incast transfer, requires certain prerequisites. That is, these flows should intersect at some network devices, thereby they can be cached and aggregated. Although the in-network aggregation of correlated flows has potential advantages in theory, the existing traffic management model is not able to realize such an outstanding design. The reason is that many data flows of an incast cannot meet at some network devices as early as possible to execute the in-network aggregation. Fortunately, with the data center topology and the locations of all sources and destinations, we can construct a cooperative transmission tree with the lowest transmission cost. With such a transmission tree, all correlated flows would transfer along the designed tree and consume the least amount of bandwidth resource.

3. *Cooperative transmission of uncertain correlated flows* The generation and propagation of east-west traffic inside a data center is often closely related to the allocation scheme of involved computing and storage resources. For performing a given job, different allocation schemes will generate entirely different flow sets. The prior cooperative transmission of correlated flows does not consider this inherent factor and only focus on realizing the in-network aggregation after flows have been generated. However, when the resource allocation schemes are not determined, the sources and destinations of the flows generated by the same job are still uncertain. Therefore, it is essential to investigate how to solve the problem of cooperative transmission of uncertain correlated flows by jointly optimizing the computing and network resources, and the storage and network resources.

References

1. Data Center [EB/OL]. [2016-01-18]. https://en.wikipedia.org/wiki/Data_center.
2. Ghemawat S, Gobioff H, Leung S T. The Google file system [C]. In Proc. of 19th ACM SOSP, New York, USA, 2003, 29–43.

3. Dean J, Ghemawat S. MapReduce: simplified data processing on large clusters [J]. Communi-
    cations of the ACM, 2008, 51(1): 107–113.
4. Chang F, Dean J, Ghemawat S, et al. Bigtable: A distributed storage system for structured data
    [J]. ACM Transactions on Computer Systems (TOCS), 2008, 26(2): 4.
5. CCF Committee on Academic Affairs, Report on Development of Computer Science and
    Technology in China [M]. Beijing: Tsinghua University Press, 2009.
6. Juve G, Deelman E, Vahi K, et al. Scientific workflow applications on Amazon EC2.
7. Sempolinski P, Thain D. A Comparison and Critique of Eucalyptus, OpenNebula and Nimbus
    [C]. In Proc. of 2nd IEEE CloudCom, Indianapolis, USA, 2010: 417–426.
8. Nurmi D, Wolski R, Grzegorczyk C, et al. The Eucalyptus Open-Source Cloud-Computing
    System [J]. Cloud Computing & Its Applications, 2009: 124–131.
9. Sotomayor B, Keahey K, Foster I. Combining batch execution and leasing using virtual
    machines [C]. In Proc. of 17th ACM HPDC, Boston, USA, 2008: 87–96.
10. Krishnan S P T, Gonzalez J L U. Building Your Next Big Thing with Google Cloud Platform
    [M]. Apress, 2015.
11. Chappell D. Introducing the windows azure platform [J]. David Chappell & Associates White
    Paper, 2010.
12. Liu Y. Introduction to the Internet of Things [M]. Beijing: Science Press, 2010.
13. [EB/OL]. [2016-01-18]. https://cloud.google.com/why-google/#support.
14. [EB/OL]. [2016-01-18]. https://aws.amazon.com/cn/.
15. Shvachko K, Kuang H, Radia S, et al. The Hadoop Distributed File System [C]. In Proc. of
    26th IEEE MSST, Nevada, USA, 2010: 1–10.
16. Jin H, Lbrahim S, Bell T, Qi L, et al. Tools and Technologies for Building Clouds [J]. Computer
    Communications & Networks, 2010: 3–20.
17. Grossman R, Gu Y. Data mining using high performance data clouds: experimental studies
    using sector and sphere [C]. In Proc. of 14th ACM SIGKDD, Las Vegas, USA, 2008: 920–927.
18. Isard M, Budiu M, Yu Y et al., Dryad: distributed data-parallel programs from sequential
    building blocks [C]. In Proc. of 23th ACM SOSP, WA, USA, 2007, 41(3): 59–72.
19. Peng D, Dabek F. Large-scale Incremental Processing Using Distributed Transactions and
    Notifications [C]. In Proc. of 11th Usenix OSDI, Vancouver, Canada, 2010: 4–6.
20. Malewicz G, Austern M, Bik A, et al., Pregel: A System for Large-scale Graph Processing [C].
    In Proc. of ACM SIGMOD, Indianapolis, USA, 2010: 135–146.
21. Melnik S, Gubarey A, Long J, et al. Dremel: Interactive Analysis of Web-scale Datasets [J].
    Comunications of the ACM, 2011, 54(6): 114–123.
22. Yu Y, Isard M, Fetterly D, et al. DryadLINQ: A System for General-purpose Distributed Data-
    parallel Computing Using a High-level Language [C]. In Proc. of 9th Usenix OSDI, San Diego,
    USA, 2008: 1–14.
23. Zaharia M, Chowdhury M, Franklin M J, et al. Spark: Cluster Computing with Working Sets
    [J]. HotCloud, 2010, 15(1):1765–1773.
24. Low Y, Gonzalez J, Kyrola A, et al. GraphLab: A New Framework for Parallel Machine
    Learning [J]. Eprint Arxiv, 2014.
25. Toshniwal A, Taneja S, Shukla A, et al. Storm@twitter [C]. In Proc. of ACM SIGMOD,
    Snowbird, USA, 2014: 147–156.
26. Buck J, Watkins N, Lefevre J. SciHadoop: Array-based Query Processing in Hadoop [C]. In
    Proc. of 25th ACM SC, Seattle, USA: ACM, 2011: 1–11.
27. Wang F, Lee R, Liu Q, et al. Hadoop-gis: A High Performance Query System for Analytical
    Medical Imaging with MapReduce. Technical report, Emory University, Aug 2011.
28. Xie J, Guo D, Hu Z, et al. Control Plane of Software-defined Networks: A Survey [J]. Computer
    Communications, 2015, 67: 1–10.

Chapter 2
State-of-the-Art DCN Topologies

Abstract The basic design goal of a DCN is to interconnect massive servers and devices with specific network topology, thereby achieving a comprehensive advantage in terms of networked computing and networked storage. The network topology plays a critical role in the DCN performance. Therefore, this chapter summarizes the latest DCN topologies and compares them in terms of construction rules, routing algorithms, network performance, and beyond. Moreover, we propose a new taxonomy that divides current DCN topologies into five categories: switch-centric, server-centric, modular, random, and wireless topologies. Finally, we outline the evolution and future trends of DCN topology designs.

2.1 Introduction

Data centers have become the core infrastructure for countries and enterprises, and their network topologies are the primary factor affecting their performance. Currently, servers are capable of forwarding packets, and switches can execute in-network computing and storage jobs. As a result, besides communication protocols among devices, the literature also focuses on network topologies, performance optimization, resource management, and energy control, with the aim of improving the comprehensive strength of computing, storage, and communication. The current development of cloud computing and big data processing faces a series of infrastructural constraints. These constraints also bring new theoretical and methodological requirements for data center design, as listed below:

1. Horizontal scalability. Traditional data centers mainly scale up in a vertical manner, i.e., increasing the number of switch ports or increasing port speed [1] directly. However, this expansion method has a considerably high cost for a large-scale data center. Therefore, various horizontal expansion methods must be explored.
2. High bandwidth. As the basic infrastructures for massive services and applications, data centers aim to satisfy enormous demand from users for computing, storage, and other resources. Aggregated network bandwidth is always a scarce resource in large data centers that directly affects the overall performance of data center applications [2].

© Springer Nature Singapore Pte Ltd. 2022
D. Guo, *Data Center Networking*,
https://doi.org/10.1007/978-981-16-9368-7_2

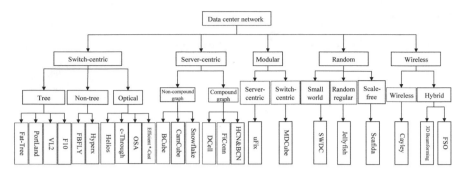

Fig. 2.1 Classification of existing data center network topologies

3. Fault tolerance. Hardware and software failures are common in large data centers [3, 4]. Data centers must ensure the security of their data against all kinds of faults, eliminate the causal impacts instantly, and provide normal and reliable service to users.

4. Cost saving and energy control. Modern data centers are incrementally replacing their dedicated devices with commodity ones to reduce hardware investment and improve the cost-benefit ratio. Growing data center sizes leads to increased energy consumption, operation/maintenance costs, and carbon emissions. Nearly 15% of the total investment of a data center goes to energy supply [5]. It is widely agreed in both academia and industry that green data centers are of great significance for future design [6].

Nowadays, there are dozens of novel DCN topology designs to meet diverse needs, with various interconnection rules and routing algorithms attached. Thus, it is necessary to classify the current topologies and summarize their evolution trend. A common classification method is to divide the topologies into switch-centric and server-centric topologies based on the forwarding and routing devices. In this chapter, we present a more fine-grained taxonomy including five categories, which cover construction rules and communication techniques. This taxonomy analyzes the topological characteristics and design philosophy of current DCN topologies to inspire new inventions.

In this chapter, existing data center networks are summarized by our taxonomy with the following five categories, as shown in Fig. 2.1: switch-centric, server-centric, modular, random, and wireless topologies. Switch-centric topologies rely on switches to realize interconnection and routing. By contrast, server-centric topologies allow servers with special network interface cards (NICs) to enable in-network interconnection and routing. Modular data centers integrate thousands of servers as container modules with cooling, maintenance, transmission, and other functions, and these modules are interconnected as large data centers. Random data centers introduce a small number of random links to interconnect remote servers or switches, thus reducing the network diameter. Extremely high-frequency (60 GHz) and laser communication technology can provide high data transmission rates [7, 8], and as such, they are implemented in wireless data center networks. Such networks can not only reduce the wiring cost on a large scale but also complement and improve the performance of wired data center networks. After detailing the above five types of DCN topology, this chapter concludes by outlining the further development trend of DCN topology design.

2.2 Switch-Centric DCN Topologies

Switch-centric network topologies can be further divided into three types: tree, flat, and optical switching topologies.

Traditional tree topologies are cascaded by three layers of switches: edge, aggregation, and core switches. These topologies incur a performance bottleneck and single-point failure at the core level. For this reason, Fat-Tree [9] is proposed to realize non-blocking transmission and eliminate the single-point failure problem. To support virtualization technologies in data centers and increase network flexibility, PortLand [10] and VL2 [11] modify the protocols on switches and servers, respectively, to enable virtual machine migration. At the same time, some works further upgrade and redesign the Fat-Tree topology. Wang et al. introduce additional aggregation switches and core switches to connect with edge switches to ease the bandwidth bottleneck problem [12]. Besides, F10 [13] modifies the symmetry characteristic of Fat-Tree to enhance its fault tolerance.

Fat-Tree and other tree-like topologies generally adopt a hierarchical interconnection structure, and the network devices at each layer are often heterogeneous. In contrast, FBFLY [14] and HyperX [15] are based on the idea of flat interconnection of homogeneous switches. Both of them adopt a generalized hypercube structure. FBFLY focuses more on reducing the overall energy consumption, while HyperX targets the optimal topology with given network resources.

Optical switching technology is applied when designing a DCN topology due to its unique advantages, including high speed, stability, and security. Helios [16] and c-Trough [17] introduce optical switches at their core layer and edge layer, respectively, so that optical circuit links and electrical packet links coexist in the network. OSA [18] discards the electrical packet switches. Instead, it introduces the Optical Switching Matrix and the Wavelength Selective Switch devices. Electrical signals are converted to optical signals on the top of the rack for transmission. Such optical signals are finally restored to electrical signals at the destination rack. The above three network topologies only introduce optical switching equipment as acceleration devices into existing networks. On the contrary, Efficient *-Cast [19], is a fully optical DCN design. It supports different traffic modes by reorganizing the optical switching devices in a dynamic and on-demand manner.

2.2.1   Tree-Like Topologies

1. Fat-Tree

Fat-Tree [9] uses a three-layer structure for switch interconnection, forming a
Clos network structure to support non-blocking communications. In Fat-Tree, edge
switches and aggregation switches are divided into different Pods. In a single Pod,
each edge switch connects to all aggregation switches, forming a complete bipartite
graph. Meanwhile, each aggregation switch is connected with some core switches.
To achieve non-blocking communication among servers, Fat-Tree deploys a large
number of aggregation switches and core switches. Assuming each switch has k
ports, Fat-Tree organizes the edge switches and aggregation switches as k Pods.
Each Pod contains $k/2$ switches at both the edge layer and aggregation layer. Such
switches use $k/2$ ports to connect devices at each of their upper and lower layers.
These Pods interconnect with the core switches to form a fully-connected bipartite
graph. Therefore, Fat-Tree requires $k^2/4$ core switches, $k^2/2$ edge switches, and $k^2/2$
aggregation switches in total.

Fat-Tree uses cheap commodity switches instead of expensive dedicated switches.
It also achieves equal aggregation bandwidth between all layers to avoid the bandwidth bottleneck problems caused by unbalanced aggregation bandwidth in multi-
root trees. Meanwhile, because there are adequate switches at the aggregation layer,
the bandwidth bottleneck problem and single-point failure problem at the aggregation layer can be resolved. Figure 2.2 shows an illustrative example of the Fat-Tree
topology for $k = 4$.

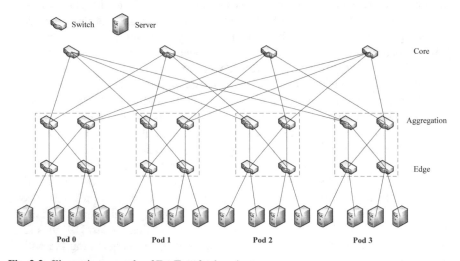

Fig. 2.2 Illustrative example of Fat-Tree for $k = 4$

2. Variants of Fat-Tree

Fat-Tree is currently the most widely applied network topologies in data centers. Many researchers have optimized and redesigned Fat-Tree to further enhance its network performance.

Wang et al. designed a new DCN topology [12] based on Fat-Tree. This variant consists of a large number of homogeneous programmable switches, and the intermediate servers divide the whole network into two Fat-Tree replicas. Each Fat-Tree replica contains a core layer, aggregation layer, and edge layer switches. As a result, every server NIC can communicate at its maximum transmission rate without being constrained by network bandwidth bottlenecks. The number of servers that this topology can accommodate depends on the number of ports in the homogeneous switches. Consequently, there are multiple paths between any pair of servers, providing high connectivity and throughput. However, it uses more switches than Fat-Tree, and the required investment is non-trivial.

In addition, Liu et al. proposed a new topology called F10, which breaks the symmetry of Fat-Tree to increase the number of paths. These paths can bypass the failure nodes when necessary [13]. F10 has three levels of fault handling mechanism: (1) bypassing the failure nodes instantly with a local rerouting mechanism, (2) switches acknowledge the paths that bypass failure nodes by pushback notification and then optimize the transmission of local traffic, and (3) continuous concurrent failures destroy the load-balance of the whole tree structure. Therefore, F10 uses a centralized scheduler to schedule traffic globally. Long-term streams are assigned with links according to their sizes. Short-term streams are routed by weighted equal-cost multi-path (ECMP) protocol. Meanwhile, in the F10 framework, nodes report failure events proactively rather than implement the heartbeat detection mechanism. In summary, F10 jointly considers routing, failure detection, and more.

3. PortLand and VL2

Compared with Ethernet, data center networks have more abundant bandwidth resources. Therefore, the layer 2 or layer 3 network protocols that are suitable for Ethernet may not be applicable in DCNs. These protocols incur the problems of scalability, communication flexibility, and virtual machine migration in DCNs. Based on the analysis of DCN topologies and network virtualization requirements, PortLand [10] has a layer 2 routing protocol with high scalability and fault tolerance. For network agility and efficiency, VL2 [12] ensures that data centers support dynamic resource allocation at the server pool level. To this end, VL2 implements the following designs: (1) the service can be placed anywhere across the network with the flat addressing scheme, (2) traffic is distributed uniformly to paths according to the load balance strategy, (3) a large-scale virtual server pool is formed based on the capability of address resolution in each end system so that the network complexity will not affect the network control plane.

PortLand is based on Fat-Tree with additional network protocol augmentation. Specifically, its routing protocol relies on its hierarchical pseudo MAC address (PMAC). By mapping the host MAC to PMAC, PortLand does not need to modify the original server configurations. In addition, PortLand uses centralized control to achieve ARP and routing fault tolerance. Meanwhile, it designs a location-based routing protocol with the symmetry characteristic of Fat-Tree. Therefore, switches automatically configure their addresses according to the location information without any human involvement. With such designs, PortLand supports virtual machine migration seamlessly.

VL2 adopts the Clos structure to provide rich link diversity and high bisection bandwidth. The VLB routing method in VL2 selects each path randomly to ensure the load balance of the network. In VL2, two sets of IP addresses are used. Network infrastructures (all switches and interfaces) use location-specific IP addresses (LAs) with location information. The link-state routing protocol only propagates LAs to ensure that switches know the switch-level interconnection structure. By contrast, external applications use fixed application-specific IP addresses (AAs). The mapping relationship between LAs and AAs is maintained by a dedicated directory system. On top of the Clos topology, the proposed VLB routing, IP address systems, and directory system jointly virtualize data centers into a huge layer-two network that supports virtual machine migration. Compared with PortLand, the disadvantage of VL2 is the need to configure agents and servers. The advantage is that if a request from a source server to a destination server is rejected, the directory server can refuse to provide an LA. In this manner, VL2 naturally enforces access control [20].

2.2.2 Flat Topologies

As a typical tree-like topology, Fat-Tree still incurs many challenges. First, Fat-Tree fails to satisfy the scalability requirement of large-scale data centers. Second, the expansion cost of Fat-Tree is high because many aggregation switches and core switches must be added. Finally, the fault tolerance of Fat-Tree is not sufficient. MSRA's research showed that Fat-Tree is very sensitive to low-level switch failures [5]. Therefore, researchers have discarded the traditional three-layer tree-like structure and constructed the flat network structure based on generic high-end homogeneous switches. FBFLY [14] and HyperX [15], for example, interconnects switches as a generalized hypercube, achieving outstanding scalability, bandwidth, and fault tolerance.

Given N nodes, the generalized hypercube [21] topology can be constructed according to the following rules. The identifier of each node is expressed with n-dimensional tuples as $X = x_n \ldots x_i \ldots x_1$, where $0 < x_i < k$ for arbitrary i. Thus, the identifiers are no longer the same as the binary variables used in standard cubes. Only when two identifiers differ in exactly one dimension can two corresponding nodes be interconnected by a link. Compared with the standard cube topology, the generalized hypercube has the characteristics of high fault tolerance, high connectivity, and short network diameter. FBFLY aims at designing a network topology that can effectively reduce the energy consumption of data centers. FBFLY uses a large number of homogeneous high-end switches for network interconnection. FBFLY is a k-ary n-

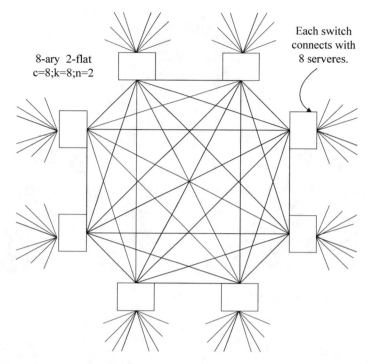

8-ary 2-flat
c=8;k=8;n=2

Each switch
connects with
8 serveres.

Fig. 2.3 Illustrative example of FBFLY

flat topology. In essence, it is a multi-dimensional generalized hypercube structure. All switches are arranged in multiple dimensions, and switches in each dimension are interconnected to form a complete graph. Figure 2.3 presents an 8-ary 2-flat FBFLY topology, wherein each node represents a 16-port switch. All 8 switches are arranged in the first dimension, and each switch connects 8 servers in the second dimension using 8 ports. Each switch uses 7 ports to connect the other 7 switches in the first dimension. Thus, there are 64 interconnected servers in total. This architecture removes the need for aggregation switches and core switches. Besides, to reduce the total energy consumption, idle network devices can be switched to their sleep state according to network traffic.

HyperX is a flattened butterfly structure and is also a generalized hypercube at its core. In the HyperX depicted in Fig. 2.4, each switch is mapped to the hypercube structure and connects to a fixed number of terminals. All switches in the same dimension are interconnected to construct a complete graph. With the given number of switch ports, bandwidth requirements, and network scale, HyperX is capable of achieving an optimal flat DCN topology.

Different from FBFLY and HyperX, which rely on a generalized hypercube to expand and organize a DCN directly, R3 [22] combines the generalized hypercube and random regular graph together to realize the joint advantages of the two types of topologies. The authors of [22] proposed a new DCN topology design methodology

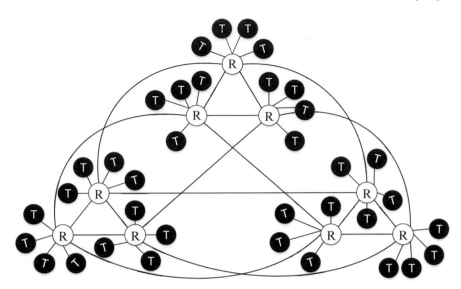

Fig. 2.4 Toy example of HyperX

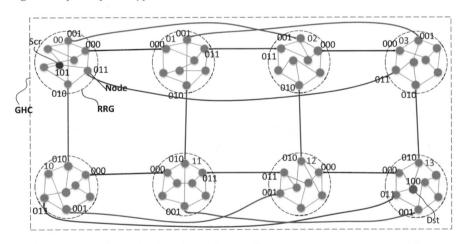

Fig. 2.5 Illustrative example of R3($G(2,4),3 - RRG$)

based on compound graph theory, which can combine the advantages of two types
of topologies without introducing their shortcomings. In particular, given a series
of basic data center modules of random topologies, a compound graph is generated
by interconnecting them into a regular topology. With such a design methodology,
different regular and random topologies can be combined to derive different hybrid
topologies. Figure 2.5 plots the hybrid structure of R3 based on a random regular
topology and hypercube topology.

With the above design, the incremental expansion characteristic of random regular topologies and easy-routing property of a generalized hypercube are effectively

combined in the generated hybrid topology. From a macro view, note that any hybrid structure can be regarded as a regular interconnection topology; from a micro view, each module acts as a random interconnection topology. In R3, the routing algorithm based on edge coloring achieves fast and accurate path search. In addition, with a given number of servers and switch ports, R3 establishes an integer programming model to optimize the topology. To achieve incremental expansion, R3 has two different expansion methods: adding new servers to basic modules and adding new basic modules to the network. Experimental results have shown that the routing flexibility and routing cost of the hybrid topology are smaller than the pure random Jellyfish structure. Moreover, the throughput of the hybrid topology is higher than that of a pure generalized hypercube.

2.2.3 Optical Switching Topologies

1. Helios and c-Through

Currently, the development of micro-electro mechanical switch (MEMS)-based optical circuit switching technology and wavelength division multiplexing technology makes it possible to construct optical switching topologies for data centers.

The Helios [16] network is a two-layer multi-root tree topology. Several servers are connected to an access switch to form a cluster. The access switch is also connected to upper-level electrical packet switches and optical circuit switches, providing both electrical and optical links. The centralized scheduler is responsible for traffic management. The Hedera [23] iterative algorithm is implemented to estimate future traffic demand. With the predication, Helios configures the network resources both dynamically and in a timely manner. Optical links are assigned to elephant flows and electrical links remain for mice flows. This allocation scheme leads to a reasonable usage of the network resources [1].

Figure 2.6 shows an example of Helios. Half of the ports in access switches are connected to hosts, and the other half are connected to the upper-layer switches. Among the ports connected to upper switches, some are equipped with optical transmitters to connect with the optical circuit switches, and the remainders are connected to the upper electrical packet switches. In this manner, the optical switching devices are embedded into the existing network seamlessly. The introduced optical links can certainly improve network performance significantly.

c-Through [17] introduces optical switching technology into its three-layer tree topology. Specifically, all edge switches are additionally connected to optical switching devices, as shown in Fig. 2.7. Therefore, c-Through can perform both packet switching and optical switching, and the decision is made by the centralized controller based on statistical traffic estimation. In c-Through, the traffic between racks can be transmitted either hop-by-hop through the tree or directly through the optical network.

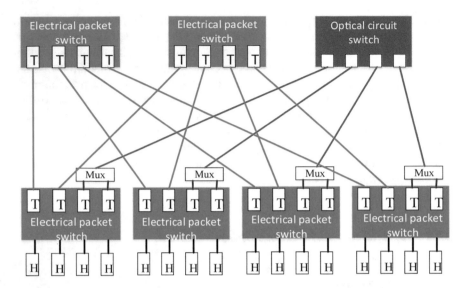

Fig. 2.6 Illustrative example of Helios

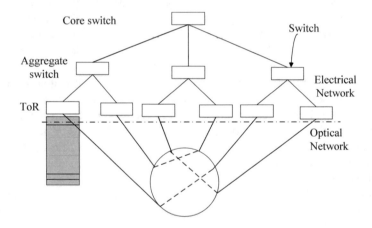

Fig. 2.7 Illustrative example of c-Through

2. OSA

Helios and c-Trough are hybrid optical network topologies, while OSA [18] is a fully optical network structure. OSA abandons the traditional tree-like topology and uses optical signal transmission throughout the entire network. As shown in Fig. 2.8, an optical transceiver is installed on each rack for photoelectric conversion. The optical signals emitted by a rack can be transmitted by one single fiber through the multiplexer (MUX). The signals with diverse wavelengths are mapped to different ports through the wavelength selective switch (WSS). Optical circulators are added for full duplex communication, and an optical switching matrix (MEMS in OSA) is

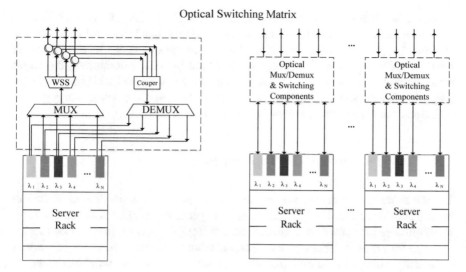

Fig. 2.8 Illustrative example of OSA

deployed to realize data exchange among different ports. Conversely, when receiving signals from other racks, the signals in multiple MEMS ports are integrated by the optical looper to the coupler, and the DEMUX re-divides the signals into multiple channels before transmission to the rack.

c-Through and Helios can only achieve single-hop optical signal transmission, while OSA is capable of multi-hop optical signal transmission with continuous photoelectric conversion. OSA greatly increases network flexibility and transmission rates. However, its disadvantage is that the network scale is limited by the number of MEMS ports, leading to limited scalability. In addition, photoelectric conversion is required for transmission, making OSA infeasible for delay-sensitive applications to some extent.

3. Efficient *-Cast

The east-west traffic patterns generated by data center-based applications include unicast, multicast, incast, and all-to-all transmission. How to efficiently support these traffic patterns of data-intensive applications is not properly resolved yet. Efficient *-Cast [19] installs a fixed-wavelength transceiver on ToR switches and connects them to high-radix optical space switches via wavelength division multiplexing. It also introduces optical switching devices, such as directional couplers and wavelength combiners, into the DCN. According to the traffic pattern in the data center, the centralized controller dynamically adjusts these optical switching devices to support data transmission. Similar to the three aforementioned optical DCN designs, unicast traffic can be transferred by just the MEMS. By splitting the signals into multiple signals with diverse wavelengths and transmitting them with dedicated ports, the multicast flows can be transferred accordingly. Combining the MUX and couplers together, this DCN design is able to transmit large-scale multicast flows. As for incast, a wavelength manipulator can be employed to integrate signals from different sources into a single port, while the remaining channels of the source can be used for other types of traffic. All-to-all traffic can be achieved by jointly combining the above three methods. Efficient*-Cast takes full advantage of optical switching devices, supports modular and incremental expansion, and can achieve better energy-saving than traditional electrical packet switching approaches.

2.3 Server-Centric DCN Topologies

The use of servers for data center networking and routing is mainly based on the following considerations. First, the openness and programmability of the server hardware/software provides a convenient means of flexible network function innovation and customization in data centers. Second, from a hardware perspective, servers are usually equipped with multiple network ports, which makes network expansion through servers possible. Finally, from an investment perspective, server-centric data centers mainly rely on servers for networking and routing, while switches only work as crossbars. Commodity switches are capable of such tasks, eliminating the need for the expensive high-end core and aggregation layers. The server-centric structure is usually constructed recursively, and the high-level network is interconnected by multiple low-level networks to achieve recursive expansion. We divide this type of topologies into the compound and non-compound graph topologies to reveal its topological properties.

2.3.1 Compound Graph-Based DCN Topologies

Given two regular graphs G and H (wherein all nodes have equal degree), the compound graph $G(H)$ is obtained by replacing each node in G with H and replacing each link in G with links connecting two nodes between two H. If the node degree in G equals the number of nodes in H, $G(H)$ is called a complete compound graph; otherwise, it is called an incomplete compound graph. The compound graph inherits the topological properties of H from a local view while maintaining the topological properties of G from a global view. A compound graph-based data center ensures high scalability of the network and high fault tolerance but falls short of achieving incremental expansion.

1. DCell

DCells are constructed iteratively and interconnected strictly using complete compound graphs during each iteration. Each layer of DCell is connected by several lower-level DCells to form a fully connected topology. The level-0 DCell is the basic building block, and its n servers are connected to an n-port switch. Specifically, each server in a lower-layer sub-network is required to connect to a server of other

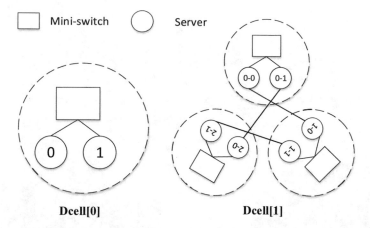

Fig. 2.9 Illustrative example of DCell

sub-networks in the same layer. More importantly, the number of lower-layer sub-networks must equal the number of servers in each lower-layer sub-network plus one [24]. If each lower-layer sub-network is abstracted as a node, the upper-layer network is a complete graph connecting these nodes. Therefore, each layer of DCell is a complete compound graph at the macroscopic level. Fig. 2.9 shows an example of DCell in which the number of switch ports is 2. If DCell[k] represents a DCell with k layers, then the number of servers N that DCell[k] can accommodate satisfies the following constraint:

$$\left(n + \frac{1}{2}\right)^{2^k} - \frac{1}{2} < N < \left(n + \frac{1}{2}\right)^{2^{k+1}} - 1 \qquad (2.1)$$

DCell has proper fault tolerance without single-point failure risk. However, as the number of layers increases, the required number of ports installed by each server also grows. One of the challenges for DCN design is how to interconnect massive servers using two network ports in each server while ensuring that the generated DCN has low network diameter and high bisection bandwidth. To this end, FiConn [25] and HCN and BCN [26] have their own solutions.

2. FiConn

Considering that current commercial servers are generally configured with at least two network ports, FiConn is a pioneering DCN topology that realizes a large-scale network with only dual network ports in each server. Similar to DCell, in FiConn, n servers connect to a n-port switch to form the basic structure directly. FiConn is also a recursively-defined network topology. For each FiConn sub-network, the second ports in half of the servers are configured to connect with servers of other sub-networks in the same layer, while the second ports in the remaining servers are reserved for building upper-layer networks.

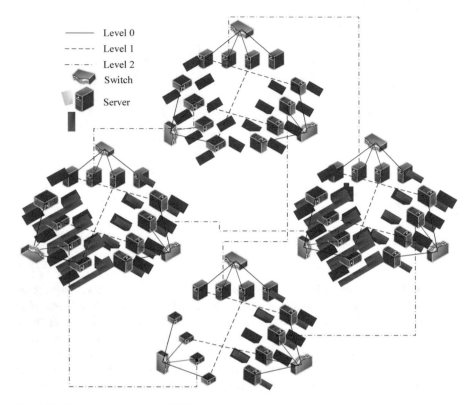

Fig. 2.10 Illustrative example of FiConn

In Fig. 2.10, if each lower-layer FiConn sub-network is regarded as a node, the upper-layer FiConn is an incomplete compound graph formed on these nodes. The number of servers that FiConn can accommodate grows exponentially with the number of ports. Its network diameter is $O(N/\log n)$ and bisection bandwidth is $O(N/\log N)$. FiConn also implements a unique traffic-aware routing algorithm to fully improve its link utilization.

3. HCN and BCN

HCN and FiConn share similar design philosophies, i.e., using a hierarchical incomplete compound graph, but they have different interconnection rules. HCN(n, h) is a h-layer network, which is constructed by embedding n HCN($n, h - 1$) sub-networks in a complete graph of n nodes according to compound graph theory. HCN($n, h - 1$) is constructed by embedding n HCN($n, h - 2$) sub-networks in a complete graph of n nodes. HCN($n, 0$) is the basic building block and is composed of n dual-port servers and one n-port switch. HCN is constructed recursively using low-end switches to provide high bandwidth and high fault tolerance while reducing the cost of network equipment. In addition, the HCN topology enjoys elegant regularity and symmetry.

Based on HCN, BCN aims to design topology with a maximum node degree of 2 and network diameter of 7 [26]. The first dimension of BCN is an incomplete compound graph defined by recursion, and the second dimension is a single-layer complete compound graph. In each dimension, the network in each layer is a complete graph of multiple sub-networks. If a 48-port switch is used, BCN with only one layer in each dimension can accommodate 787,968 servers, while 2-layer FiConn can only accommodate 361,200 servers. Neither HCN nor BCN has an upper limit on a network scale, and they can expand continuously on demand. HCN and BCN are expanded from a two-dimensional plane and multi-dimensional space respectively, forming hierarchical incomplete compound graphs from a global perspective.

2.3.2 Non-Compound Graph-Based DCN Topologies

In addition to the core compound graph concept, researchers have also employed other well-structured topologies to organize server-centric DCNs. For example, BCube [27] interconnects servers and expands indirectly in the manner of a generalized hypercube. Camcube [28] interconnects servers according to the torus topology, while Snowflake [29] relies on Koch curves to organize servers and expand recursively.

1. BCube

BCube [27] is a topology designed for modular data centers, i.e., it is a container-level topology. BCube is expanded hierarchically to interconnect a large number of servers. BCube interconnects several lower-layer BCube networks as generalized hypercubes through upper-layer switches. As shown in Fig. 2.11, each upper-layer switch is connected to all lower-layer BCube networks. For every expansion, the number of switches required at the added layer is jointly determined by the number of ports on switches and the number of layers in the network. $BCube_0$ is obtained by connecting n servers to an n-port switch. When building $BCube_1$, an additional n upper-layer switches are required. Each upper-layer switch is connected to all n $BCube_0$, thereby constructing a larger BCube network recursively. The number of servers that a k-layer BCube can hold is n^{k+1}, at the cost of $k + 1$ ports at each server. There are at most $k + 1$ parallel paths between any pair of servers in $BCube_k$. When transmitting one-to-many flows, the servers in the same layer form a complete graph with network diameter 2.

2. CamCube

The implementation of "external services" (such as search, mail, and shopping) provided by a data center is mainly grounded in the interfaces provided by many "internal services". Therefore, internal services are black boxes to external services. External services cannot know the specific circumstances of internal services. To solve this problem, CamCube [28] adopts the symbiotic routing method. Basically,

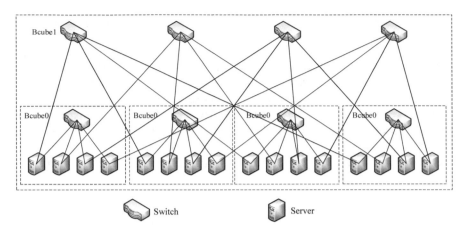

Fig. 2.11 Illustrative example of BCube

CamCube employs the 3D torus architecture. Each server is assigned with a three-dimensional coordinate to represent its logical location in the network. Based on this coordinate, a server connects with a few neighboring servers without introducing any routers or switches. In particular, in CamCube, external services can access any server in the coordinate space directly. Meanwhile, servers are allowed to eavesdrop on and modify packets. As a result, external services can implement their own customized routing protocols through servers. CamCube supports also multi-hop routing using link-state protocols.

3. Snowflake

The Snowflake structure [29] is also defined recursively. The basic unit $Snow_0$ is constructed according to a Koch curve [30] and is composed of a switch connected to three servers. In Fig. 2.12, the solid line indicates the actual physical connection, and the dashed line denotes the virtual connection (a connection that does not actually exist). When building $Snow_1$, all virtual connections are replaced by cell units ($Snow_0$), and the switch in the added cells are connected to the original server directly. The subsequent expansion follows the same approach by replacing all real and virtual connections with cells. Such a design ensures that there are more than two parallel paths between any pair of servers.

The Snowflake structure uses the network level and angle (0 degrees in the north direction and counts the degree clock-wisely) to form a binary tuple $<level, degree>$ to uniquely identify the switch and server. For instance, the tuple $<1, 120>$ indicates that the node is in the 120 degree position in $Snow_1$. Based on this binary identity, Snowflake specifies its own routing strategy. However, the Snowflake structure is not suitable to interconnect large-scale servers. When the network scale reaches an order of 10^6, the servers in $Snow_1$ must be replaced with switches, and a heartbeat mechanism is required to judge whether a switch works normally or not. These potential disadvantages increase the difficulty of expanding and maintaining the Snowflake

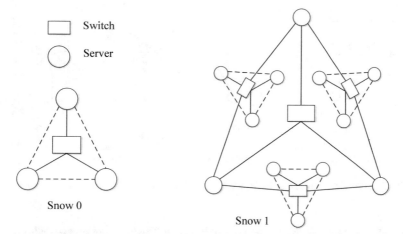

Fig. 2.12 Illustrative example of Snowflake

topology. Furthermore, the switch in the center of Snowflake will become a bottleneck of the network, and its failure may cause catastrophic damages to the whole network.

2.4   Modular DCN Topologies

There are two mainstream methodologies for constructing large-scale data centers. The first is to construct an integrated large-scale data center through scalable network topologies, such as DCell, BCube, and HCN. The second is to construct large-scale data centers by interconnecting many modules, each constituting an independent small-scale data center. With the rapid development of technologies, modular data centers have replaced racks as the basic unit to build large-scale data centers. Modular data centers have advantages in terms of instant configuration, mobility, system and energy density, cooling and configuration costs, etc. Modular DCNs have become the basic solution for efficient, controllable, manageable, plug-and-play, and flexible data centers [31]. Modular data centers involve two levels of topology: intra-module and inter-module.

2.4.1   Intra-module Network Topology

BCube is a topology designed for intra-module interconnection. DCube [32] is an alternative topology that interconnects many dual-NIC servers and cheap switches. DCube(n, k) consists of k sub-networks, each constructed by interconnecting many

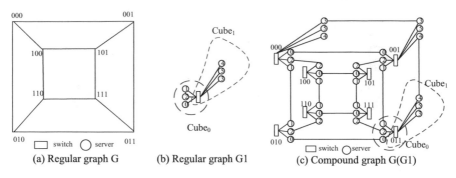

Fig. 2.13 Illustrative example of H-DCube for $n = 6, k = 2$

hypercube-like units with compound graph theory. The building unit in DCube (n, k) is composed of n/k servers and a switch. In fact, the DCube topology contains two different designs: H-DCube based on a hypercube (as shown in Fig. 2.13) and M-DCube based on a Möbius cube.

H-DCube and M-DCube are fully regular and symmetrical. Such features are very important for data center networks. The diameter of a Möbius cube is approximately half that of a standard hypercube, while the diameter of an M-DCube network is approximately 2/3 that of an H-DCube network. Compared with BCube and Fat-Tree, DCube provides better network bandwidth and fault tolerance in unicast communication. Besides, DCube requires much fewer cables and switches.

2.4.2 Inter-module Network Topology

M-DCube [33] is an inter-module topology to interconnect modules formed as BCubes. As shown in Fig. 2.14, M-DCube is essentially a multi-dimensional generalized hypercube, and the number of data center modules it can accommodate is equal to the product of the number of modules that can be accommodated in each dimension. To speed up the transmission among modules, M-DCube uses optical fibers to transmit cross-module flows. Moreover, both ends of the cross-module links use reserved 10 Gbps ports. Such links can be used when necessary. The hypercube structure in M-DCube provides multiple parallel paths for cross-module flows, so M-DCube is highly fault tolerant.

M-DCube focuses on how to interconnect homogeneous data center modules. However, a production data center may introduce heterogeneous modules at different construction periods, and it is challenging to integrate these into an existing DCN. uFix [34] presents a solution to this problem. As shown in Fig. 2.15, uFix interconnects heterogeneous modules based on incomplete compound graphs. Specifically, the across-module links connect the idle ports in these modules. The advantage of uFix is that there is no need to introduce additional network devices when intercon-

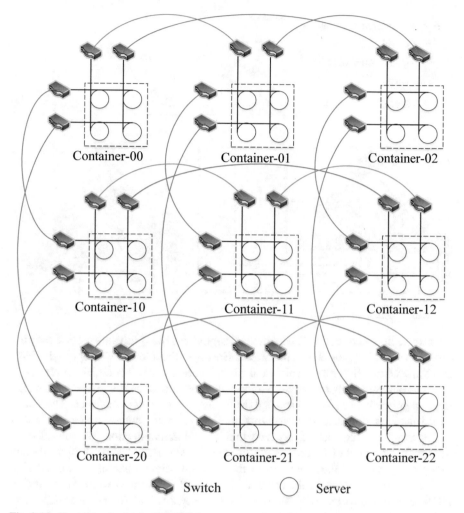

Fig. 2.14 Illustrative example of M-DCube

necting the modules. However, it fails to ensure that all modules are connected to the DCN when the DCN is too large. uFix also incurs several uncertainties. For instance, it is uncertain whether there is an idle server port in each data center module and whether the number of remaining ports in the server will limit the further expansion.

2.5 Random DCN Topologies

In the above DCN topologies, expensive dedicated switches are replaced with general commodity switches, and servers with special network ports are also employed for

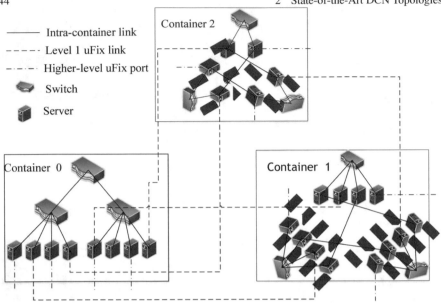

Fig. 2.15   Illustrative example of uFix

interconnection and routing. These novel designs have many advantages to satisfy the diverse needs of upper-level applications. However, the above DCN topologies have strict interconnection rules, which make it difficult to expand and maintain them.

As an emerging subject, network science has developed rapidly in recent years. ER random graph theory [35, 36] has laid the foundation of network science. The small-world network and scale-free network models are two representative milestones in ER graph theory. Having realized that strict interconnection rules restrict the further development of data center performance, the academic community has begun to seek potential breakthroughs from the network science paradigm. This design approach aims to maintain the original topological advantages while guaranteeing DCN performance in terms of bandwidth, scalability, and fault tolerance. Researchers have proposed the small-world data center (SWDC) [37] based on the small-world network model, Jellyfish [38] based on the random regular graph model, and Scafida [39] based on the scale-free network model. Among them, SWDC introduces a certain number of random links into typical regular network topologies. These links are added with a probability proportional to the distance between two network nodes, thereby reducing the network diameter. Jellyfish establishes random links among switches, ensuring incremental expansion of the DCN. Scafida is highly fault tolerant with the guarantee that the node degree is less than the number of ports.

Random DCNs abandon the restriction of interconnection rules and thereby enjoy design flexibility. However, they also bring several disadvantages, such as network capacity, wiring, and maintenance. SWDC realizes interconnections among servers without the use of switches, but the network scale is limited by the number of network ports in each server. Jellyfish must adjust some existing links when adding new switches. The node degree in Scafida follows a power law distribution. The hub node is the key of the whole network, and its failure may paralyze the whole network. Although the interconnection rules in random DCNs are relatively simple, connecting two remote nodes increases the wiring cost and difficulty. In addition, in the process of troubleshooting, it is difficult to operate and maintain the network because of its irregularity.

2.5.1 Small-World Network-Based DCN Topology

The small-world network concept [40] was first introduced by Watts and Strogatz. They introduced random connections into a regular ring topology to form the network. On this basis, Kleinberg presented a new way to build small-world networks [41]. On a 2D grid topology, random links were added between two nodes with a probability proportional to the distance between them. As a result, the average network diameter was reduced to a magnitude of $O(logN)$. There are three types of SWDC, including ring-based, 2D torus-based, and 3D torus-based.

Figure 2.16 shows a 2D torus-based SWDC topology, where the solid line represents the regular links of a torus, and the dashed line denotes the added random links. If the node degree is 6, each node will connect to 2 regular links and 4 random links in a ring-based SWDC, 4 regular links and 2 random links in a 2D torus-based SWDC, and 5 regular links and 1 random link in a 3D torus-based SWDC. The SWDC structure assigns a unique geographic identity to each server and implements a greedy algorithm to obtain the shortest routing path with both random links and regular links. The SWDC topology has the advantages of high bandwidth, low network diameter, and easy deployment.

2.5.2 Random Regular Graph-Based DCN Topology

Of the various structured topologies, the torus is often employed in supercomputers, the hypercube is often used in data centers and supercomputers, and tree topologies are often implemented in data centers. However, the network scale of structured topologies can only be extended sharply. For example, the Fat-Tree using 24-port edge switches can interconnect 3,456 servers, while the Fat-Tree using 32-port edge switches can interconnect 8,192 servers. In a production data center, the network scale is usually expanded in an incremental manner. If there is a need to expand the DCN scale from 3456 servers to 4000 servers, there are two approaches. The first is to continue to use the original 24-port switches and introduce more racks to deploy the new servers. These racks connect with the core layer switches directly. However, this approach will damage the symmetry of the original Fat-Tree, causing asymmetric bandwidth. The second approach is to replace the 24-port edge switches with 32-port switches, resulting in more idle switch ports and unnecessary investment.

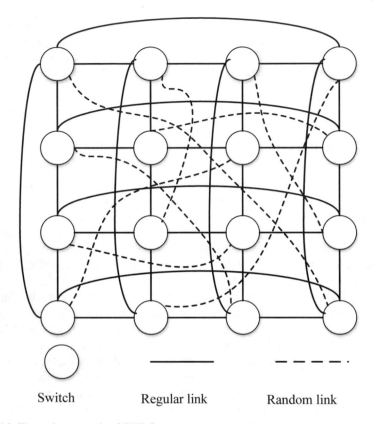

Switch   Regular link   Random link

Fig. 2.16 Illustrative example of SWDC

Jellyfish is an incremental scalable topology to solve this problem (as shown in Fig. 2.17). Jellyfish constructs random regular graphs using homogeneous switches. For each switch, some of its ports are used to connect servers, while the remaining ports connect with other switches. When adding a new switch to the existing network, a randomly selected link is removed, and the two nodes connected by the removed link are connected to the added switch. Jellyfish ensures the high utilization of network bandwidth by reducing the average path length. It employs the k-shortest path method for routing and uses the multi-path TCP protocol for congestion control. It also optimizes the placement pattern of switches.

2.5.3 Scale-Free Network-Based DCN Topology

As shown in Fig. 2.18, Scafida is constructed based on a scale-free network. The degree distribution of a scale-free network follows the power law distribution. For the majority of nodes (switches), the degree is small. By contrast, a few nodes act as

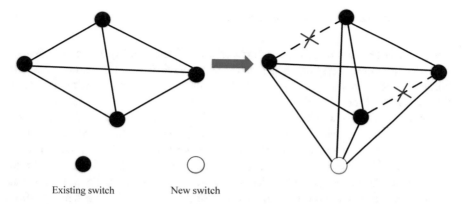

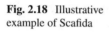

Existing switch   New switch

Fig. 2.17 Illustrative example of Jellyfish

Fig. 2.18 Illustrative example of Scafida

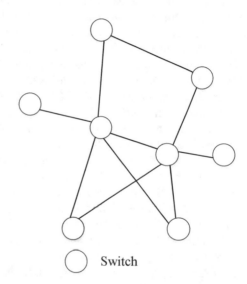

Switch

hub nodes with extremely large degrees. As a result, such a scale-free network enjoys a low network diameter. The algorithm for building a scale-free network proposed by Barabási is based on the two assumptions that the network scale continuously expands and that newly added nodes are more inclined to connect to nodes with higher degrees. Theoretically, the degree of hub nodes can be unbounded. However, in practice, node degree is limited by the number of ports in switches and servers. Note that after switches are networked, the remaining ports will be connected to the servers directly. Experimental results have shown that the limitation of node degree has a significant impact on network performance.

2.6   Wireless DCN Topologies

The above four types of data center networks inevitably deploy a large number of links in data centers, resulting in huge labor costs and difficulties in maintenance and failure detection. In addition, the congestion caused by burst traffic will still restrict the DCN performance.

Although 60 GHz wireless communication technology has many advantages, its signal transmission incurs great transmission loss, poor diffraction, and penetration, which make it only capable of short-range communication. In Cayley, the server is stacked as cylinder racks. The directional antenna can be adjusted dynamically to establish one-hop wireless communication between any servers in the same rack, but long-range communication between racks can only be achieved through hop-by-hop forwarding. 3D Beamforming improves signal intensity and transmission distance by beamforming and reflection technologies. In addition, dedicated hardware for 60 GHz communication is expensive. This problem can be alleviated to some extent by fabricating chips using CMOS, but it is still too expensive compared to wired communication equipment. Finally, when sending and receiving signals, the transceiver device is required to adjust the main lobe direction continuously. Therefore, the transceiver should be controlled precisely.

The main shortcomings of 60 GHz technology include potential signal interference and short-range transmission. Compared with 60 GHz radio frequency technology, laser communication technology has the advantages of anti-interference, low bit error rate, high bandwidth, long-distance transmission, and more [8]. However, the challenge is that the laser signals are propagated directionally, so they can easily be blocked by obstacles. Free space optics (FSO) [42] is a fully wireless inter-rack data center fabric constructed based on laser communication technologies. It adjusts the states of switchable mirrors on demand. The laser beams are reflected by a top ceiling mirror to realize inter-rack communications.

Wireless DCN is one of essential development trends for data centers. However, current technology is not compact enough to construct an ideal wireless or hybrid DCN topology to simultaneously satisfy the needs of bandwidth, fault tolerance, incremental expansion, and low cost.

2.6.1   60 GHz-Based Hybrid DCN Topology

Although 60 GHz wireless communication technology has the advantages of high bandwidth and high transmission rate, it still suffers from two problems when used in data centers. First, it can only ensure high bandwidth for short-range communication. Second, wireless signals are very sensitive to obstacles. To ensure efficient wireless communication between any two racks in a data center, Van Veen et al. proposed the beamforming technique [43] to form a narrow emission beam that concentrates energy towards the target direction. This technique enhances the transmission dis-

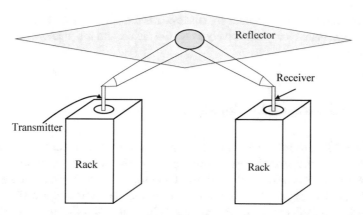

Fig. 2.19 Illustrative example of 3D Beamforming-based topology

tance of wireless signals significantly. Even so, it still falls short of achieving long-range inter-rack wireless communication. To this end, Van Veen proposed installing a reflective ceiling at the top of the data center, as shown in Fig. 2.19. By controlling the elevation of the transmitter and receiver, the wireless signal can reach the receiver exactly after being reflected by the reflective ceiling. This scheme protects the wireless signals from being blocked by obstacles, thereby achieving long-distance communication within the data center.

2.6.2 60 GHz-Based Wireless DCN Topology

A Cayley graph [44] is a highly symmetric graph constructed by the British mathematician Cayley in 1895. It has excellent symmetry and transitivity and is considered to be a proper multi-computer interconnection topology. It has been applied in P2P networks [45, 46] by researchers. When building a DCN with a Cayley graph, two wireless transceivers are each installed at the front and back ends of each server. 20 servers form one layer of circular structure, and 5 layers of such cycles form a cylinder rack. Intra-rack communication is achieved by the transceivers inside of the rack. Each transceiver can only communicate with the servers within its main lobe. Therefore, the servers at the same layer form a Cayley graph. Inter-layer communication is realized by adjusting the elevation of the pair of transceivers. Such racks are placed as a grid topology. The communication between adjacent racks is realized by a series of external transceivers. Generally, a rack can only perform one-hop wireless communication with adjacent racks, and communication with other racks must be relayed. An efficient routing protocol [47] has been designed based on the geographic routing technique.

As for the obstacles encountered along the wireless transmission path, Cayley has no solution. We believe that we can learn from the idea of 3D Beamforming and build a reflection mirror to reflect signals towards the receiving rack. The advantage of this method is that it can effectively avoid obstacles, and the placement of the rack is flexible.

2.6.3   FSO-Based DCN Topology

Similar to 3D Beamforming, FSO [42] also focuses on providing extra wireless links for inter-rack communication and forming a specific wireless network topology. These two schemes make modifications at the top of racks to install transceivers and realize out-of-sight transmission of wireless signals through a celling mirror. The differences between them include the following. (1) FSO uses lasers as the signal carriers and can support long-range communication at the kilometer level. (2) 3D Beamforming emits wireless signals towards different areas by adjusting the elevation angle of the transceivers, while FSO achieves data transmission by adjusting the state of switchable mirrors (in the "glass" state to allow the signal to pass through, while in the "mirror" state to reflect signals [48]). (3) FSO can be dynamically re-constructed as different topologies, such as random regular graph and hypercube with randomly added links. It achieves the proper topological properties of the topologies (such as network diameter and bisection bandwidth) and avoids high wiring cost simultaneously. FSO is just an exploratory scheme of how to introduce free-space optical communication technology into data centers. There are still many problems faced by this technology before its real-world application. For example, how to realize multipath transmission and manage traffic.

2.6.4   VLC-Based DCN Topology

In addition to 60 GHz wireless communication and laser communication technology, researchers have further explored how to introduce visible light communication (VLC) into the design of DCN topology. Such a design upgrades the network capability of existing wired DCNs. As shown in Fig. 2.20, VLCcube is a hybrid DCN topology consisting of both wireless and wired links. Specifically, based on Fat-Tree, VLCcube installs four VLC transceivers at the top of each rack to provide four 10 Gbps wireless links to form a wireless torus topology. With the wireless torus, VLC-cube provides shorter paths for many flows that would require 4-hop transmission in the original Fat-Tree. Therefore, VLCcube achieves better network performance than Fat-Tree, and its congestion-aware scheduling strategy further improves the performance. VLCcube is one instance of deploying VLC links in data centers. In the future, vendors can design a completely different hybrid DCN based on a different wired DCN. The introduction of VLC links functions well within existing wired DCNs and enhances the network flexibility and performance.

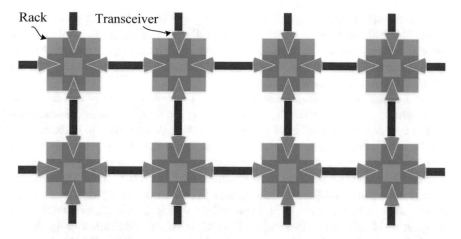

Rack   Transceiver

Fig. 2.20   Illustrative example of VLCcube

2.7   Evolution and Future Trends of DCN Topology Design

As stated earlier, many achievements have been made in the field of DCN topology design, and novel topologies are continuously emerging. Given so many designs, it is necessary to determine the evolution process of such designs to help guide the future development of DCNs.

2.7.1   Evolution of DCN Topology Design

Traditional data centers adopt the tree-like topologies, which incur potential single-point failure and bandwidth bottleneck. Fat-Tree forms edge layer switches and aggregation layer switches as different Pods and then constructs a Clos structure among the three layers of switches to enable non-blocking communication. To meet the increasing demand of virtual machine migration, PortLand and VL2 modify the protocols on switches and servers, respectively, to support virtual machine migration without IP address changes. Such modifications also improve the DCN flexibility. Based on Fat-Tree, F10 constructs an asymmetric AB Fat-Tree to increase the number of paths bypassing failure nodes in the network. Based on its failure report mechanism, F10 provides three levels of failure recovery strategy (local, partial, and global) to guarantee instant failure recovery.

FBFLY and HyperX discard the hierarchical tree structure. Instead, FBFLY uses a flat structure of a generalized hypercube to reduce the number of switches and adjusts links dynamically to save energy. HyperX aims to realize an optimal topology with given network requirements and resource constraints. The rise of optical circuit switching technology has a great impact on DCN topology design. Helios and c-

Through introduce optical switches to existing tree topologies and provide both optical fiber links and packet switching links. OSA aims to construct a fully optical network. However, frequent photic-electric conversions are required for multi-hop transmission in OSA. Efficient *-Cast handles different types of traffic through the dynamic reorganization of optical devices.

The continuous upgrading of servers has led to the emergence of server-centric DCN topologies. These are usually constructed recursively, i.e., the upper-level network is expanded by a number of lower-layer networks according to certain rules. This kind of topology aim to construct an excellent data center network with a large number of generic servers and commodity network devices. They jointly consider multiple features, such as link utilization, load balance, efficient routing, and on-demand expansion. Modular data centers follow a completely different design methodology. In such centers, all basic facilities are integrated into a module to build a large-scale DCN quickly. In addition, SWDC, Jellyfish, and Scafida apply classic models from network science to construct DCN topologies with low network diameter, high fault tolerance and strong scalability. However, such irregular topologies lead to high routing complexity and increased maintenance costs.

Emerging wireless communication technologies enable novel wireless DCN ideas. 3D Beamforming uses beamforming technology to enhance signals and constructs reflectors to bypass the barriers between racks. Cayley is a fully wireless network built on the basis of altered servers. Similarly, FSO provides path diversity of laser signals by adjusting switchable mirrors.

For current DCNs, the design method has evolved from regular to random, from wired to wireless, from static to dynamic, and from integrated to modular. Such changes are closely related to the development of communication technologies. It is foreseeable that future new technologies for equipment and communication will provide additional opportunities for DCN topology design improvement.

2.7.2   Future Trends of DCN Topology Design

With the development of core equipment and communication technologies in data centers, future DCN topology design has multiple opportunities as well as challenges.

First, there are heterogeneous and incrementally scalable topologies. Most current DCN designs concern how to interconnect servers and switches of the same type or the same number of ports. As a matter of fact, the construction and expansion of data centers will inevitably introduce heterogeneous computing and networking devices. Therefore, it is necessary to study how to effectively organize and expand these heterogeneous devices. At present, computing equipment and networking devices tend to be integrated in a single device. Meanwhile, technologies of wired, optical, and wireless communication for data centers are constantly emerging. These potential developments bring great challenges to the design of heterogeneous and incrementally scalable DCNs.

Second, there are modular data centers. A module contains a small number of servers organized as a regular topology. Such modules are further interconnected as a large-scale DCN. This design philosophy also incurs the problem of heterogeneous expansion, i.e., how to expand network scale with diverse equipment introduced at different times. Although uFix is devoted to solving this problem, further investigation is still required. In the future, cloud providers can use optical circuit switching or wireless communication technology to interconnect data center modules for better scalability and flexibility.

Third, there are wireless DCN topologies. The transmission of data with better transmission rate and bandwidth using light as the carrier is ubiquitous. Laser communication provides abundant bandwidth and spectrum resources. The emerging VLC (Li-Fi) uses the pulse signal emitted by an LED [49] to realize signal transmission (approximately 10 Gbps). Such VLC techniques enjoy low-cost investment and elegant controllability. In the future, there will be more kinds of optical communication devices and high-performance optical network components. These devices and components will accelerate the development of fully optical DCNs. Optical DCN and other wireless DCN designs are still in their experimental stages, and further improvement and designs are required before they enter the market.

Fourth, there are software-defined data center networks. An SDN [50] decouples the network control plane from physical devices. Such control functionalities are integrated into the centralized controller, thereby avoiding the restrictions of physical devices. A software-defined data center (SDDC) [51] is a combination of an SDN and DCN, which extends the design space of DCN topologies. An SDDC virtualizes all infrastructures and implements Infrastructure as a Service (IaaS) through intelligent software to schedule resources dynamically. Traffic control, load balancing, routing protocols, controller placement, customizable protocols, fault tolerance, disaster tolerance, resource allocation, network security, and other issues in SDDCs deserve further study [52–54]. For example, in a server-centric data center network, servers provide networking and routing functions, so the SDN controller required further redesign. In addition, optical switching equipment, wireless switching devices, server NIC cards, and traditional networking modules coexist in data centers. Such a phenomenon makes Openflow-based SDN technology no longer applicable. Consequently, it is urgent to design new SDN solutions.

References

1. Li D, Chen G, Ren F, et al. Data Center Network Research Progress and Trends [J]. Chinese Journal of Computers, 2014, 37(2): 259–274.
2. Prasad R, Dovrolis C, Murray M, et al. Bandwidth estimation: metrics, measurement techniques, and tools [J]. IEEE Network, 2003, 17(6): 27–35.
3. Maltz D A. Challenges in cloud scale data centers [C]. In Proc. of ACM SIGMETRICS, Pittsburgh, USA, 2013: 3–4.
4. Wu X, Turner D, Chen C C, et al. Netpilot: automating datacenter network failure mitigation [J]. ACM SIGCOMM Computer Communication Review, 2012, 42(4): 419–430.

5. Greenberg A, Hamilton J, Maltz D A, et al. The cost of a cloud: research problems in data center networks [J]. ACM SIGCOMM Computer Communication Review, 2008, 39(1): 68–73.
6. Bostoen T, Mullender S, Berbers Y. Power-reduction techniques for data-center storage systems [J]. ACM Computing Surveys (CSUR), 2013, 45(3): 33.
7. Ranachandran K, Kokku R, Mahindra R, et al. 60 GHz data-center networking: wireless=> worryless [J]. NEC Laboratories America, Inc., Tech. Rep., 2008.
8. Kedar D, Arnon S. Urban optical wireless communication networks: the main challenges and possible solutions [J]. Communications Magazine, IEEE, 2004, 42(5): S2–S7.
9. Al-Fares M, Loukissas A, Vahdat A. A scalable, commodity data center network architecture [J]. ACM SIGCOMM Computer Communication Review, 2008, 38(4): 63–74.
10. Niranjan Mysore R, Pamboris A, Farrington N, et al. Portland: a scalable fault-tolerant layer 2 data center network fabric [J]. ACM SIGCOMM Computer Communication Review, 2009, 39(4): 39–50.
11. Greenberg A, Hamilton J R, Jain N, et al. VL2: a scalable and flexible data center network [J]. ACM SIGCOMM Computer Communication Review, 2009, 39(4):51–62.
12. Wang C, Wang C, Wang X, et al. Data Center Network Architecture Design towards Cloud Computing [J]. Computer Research and Development, 2012, 49(2): 286–293.
13. Liu V, Halperin D, Krishnamurthy A, et al. F10: A fault-tolerant engineered network [C]. In Proc. of 10th NSDI, Lombard, USA, 2013: 399–412.
14. Abts D, Marty M R, Wells P M, et al. Energy proportional datacenter networks [J]. ACM SIGARCH Computer Architecture News, 2010, 38(3): 338–347.
15. Ahn J H, Binkert N, Davis A, et al. HyperX: topology, routing, and packaging of efficient large-scale networks [C]. In Proc. of SC, New York, USA, 2009: 41.
16. Farrington N, Porter G, Radhakrishnan S, et al. Helios: a hybrid electrical/optical switch architecture for modular data centers [J]. ACM SIGCOMM Computer Communication Review, 2011, 41(4): 339–350.
17. Wang G, Andersen D G, Kaminsky M, et al. C-Through: Part-Time Optics in Data Centers. ACM SIGCOMM Computer Communication Review [J]. 2010, 40(4): 327–338.
18. Chen K, Singla A, Singh A, et al. OSA: an optical switching architecture for data center networks with unprecedented flexibility [J]. IEEE/ACM Transactions on Networking, 2014, 22(2): 498–511.
19. Wang H, Xia Y, Bergman K, et al. Rethinking the physical layer of data center networks of the next decade: using optics to enable efficient*-cast connectivity [J]. ACM SIGCOMM Computer Communication Review, 2013, 43(3): 52–58.
20. It's Microsoft vs. the professors with competing data center architectures [EB/OL]. [2016-01-18]. http://www.networkworld.com/news/2009/082009-microsoft-sigcomm.html?page=2.
21. Bhuyan L N, Agrawal D P. Generalized hypercube and hyperbus structures for a computer network [J]. IEEE Transactions on Computer, 1984, 100(4): 323–333.
22. Luo L, Guo D, Li W, et al. Compound graph based hybrid data center topologies [J]. Frontiers of Computer Science, 2015, 9(6): 860–874.
23. Al-Fares M, Radhakrishnan S, Raghavan B, et al. Hedera: Dynamic Flow Scheduling for Data Center Networks [C]. In Proc. of 7th USENIX NSDI, San Jose, USA, 2010.
24. Guo C, Wu H, Tan K, et al. Dcell: a scalable and fault-tolerant network structure for data centers [J]. ACM SIGCOMM Computer Communication Review, 2008, 38(4): 75–86.
25. Li D, Guo C, Wu H, et al. FiConn: Using backup port for server interconnection in data centers [C]. In Proc. of 28th IEEE INFOCOM, Rio de Janeiro, Brazil, 2009: 2276–2285.
26. Guo D, Chen T, Li D, et al. Expandable and cost-effective network structures for data centers using dual-port servers [J]. IEEE Transactions on Computers, 2013, 62(7): 1303–1317.
27. Guo C, Lu G, Li D, et al. BCube: a high performance, server-centric network architecture for modular data centers [J]. ACM SIGCOMM Computer Communication Review, 2009, 39(4): 63–74.
28. Abu-Libdeh H, Costa P, Rowstron A, et al. Symbiotic routing in future data centers [J]. ACM SIGCOMM Computer Communication Review, 2011, 41(4): 51–62.

29. Liu X, Yang S, Guo L, et al. Snowflake: a new-type network structure of data center [J]. Chinese Journal of Computers, 2011, 34(1): 76–86.
30. Lapidus M L, Pearse E P J. A tube formula for the Koch snowflake curve, with applications to complex dimensions [J]. Journal of the London Mathematical Society, 2006, 74(02): 397–414.
31. Hamilton J. Architecture for modular data centers [C]. In Proc. of 3th CIDR, California, USA, 2007.
32. Guo D, Li C, Wu J. DCube: A family of network structures for containerized data centers using dual-port servers [J]. Computer Communications, 2014, 53: 13–25.
33. Wu H, Lu G, Li D, et al. MDCube: a high performance network structure for modular data center interconnection [C]. In Proc. of 5th ACM CoNEXT, Rome, Italy, 2009: 25–36.
34. Li D, Xu M, Zhao H, et al. Building mega data center from heterogeneous containers [C]. In Proc. of 19th IEEE ICNP, Vancouver, Canada, 2011: 256–265.
35. Erdos P, Renyi A. On random graphs. Publ. Math. Debrecen, 1959, 6: 290–297.
36. Erdos P, Renyi A. On the evolution of random graphs [J]. Publ. Math. Inst. Hung. Acad. Sci, 1976, 2: 482–525.
37. Shin J Y, Wong B, Sirer E G. Small-world datacenters [C]. In Proc of 2nd ACM SOCC, Cascais, Portugal, 2011: 1–13.
38. Singla A, Hong C Y, Popa L, et al. Jellyfish: Networking data centers randomly [C]. In Proc. of 9th USENIX NSDI, San Jose, USA, 2012: 17–17.
39. Barabási A L, Albert R. Emergence of scaling in random networks [J]. Science, 1999, 286(5439): 509–512.
40. Watts D J, Strogatz S H. Collective dynamics of small-world networks [J]. Nature, 1998, 393(6684): 440–442.
41. Kleinberg J. The small-world phenomenon: An algorithmic perspective [J]. In Proc. of 32th STOC, Portland, USA, 2000: 163–170.
42. Hamedazimi N, Gupta H, Sekar V, et al. Patch panels in the sky: a case for free-space optics in data centers [C]. In Proc. of 12th ACM SIGCOMM Workshop on Hot Topics in Networks (HotNets), Hong Kong, China, 2013: 1–7.
43. Van Veen B D, Buckley K M. Beamforming: A versatile approach to spatial filtering [J]. IEEE ASSP Magazine, 1988, 5(2): 4–24.
44. Sylvester J J. On an application of the new atomic theory to the graphical representation of the invariants and covariants of binary quantics, with three appendices [J]. American Journal of Mathematics, 1878, 1(1): 64–104.
45. Peng L. Research on wireless P2P overlay model and key technologies based on Cayley graphs [D]. South China University of Technology, 2011.
46. Liang H. Topology Construction and Resource Locating of Structured P2P Overlay Network Based Cayley Graph [D], South China University of Technology, 2012.
47. Shin J Y, Sirer E G, Weatherspoon H, et al. On the feasibility of completely wirelesss datacenters [J]. IEEE/ACM Transactions on Networking (TON), 2013, 21(5): 1666–1679.
48. Switchable Mirror/Switchable Glass [EB/OL]. [2016-01-18]. http://kentoptronics.com/switchable.html.
49. Burchardt H, Serafimovski N, Tsonev D, et al. VLC: Beyond point-to-point communication [J]. IEEE Communications Magazine, 2014, 52(7): 98–105.
50. McKeown N, Anderson T, Balakrishnan H, et al. OpenFlow: enabling innovation in campus networks [J]. ACM SIGCOMM Computer Communication Review, 2008, 38(2): 69–74.
51. The Software-Definedd-Data-Center (SDDC): Concept Or Reality [EB/OL]. [2016-01-18]. http://blogs.softchoice.com/advisor/ssn/the-software-defined-data-center-sddc-concept-or-reality-vmware/.
52. Jia W-K. A Scalable Multicast Source Routing Architecture for Data Center Networks [J]. IEEE Journal on Selected Areas in Communications, 2014, 32(1):116–123.
53. Lester A, Tang Y, Gyires T. Prioritized Adaptive Max-Min Fair Residual Bandwidth Allocation for Software-Defined Data Center Networks [C]. In Proc. of 13th ICN, Nice, France, 2014: 198–203.

54. Szyrkowiec T, Autenrieth A, Gunning P, et al. First field demonstration of cloud datacenter workflow automation employing dynamic optical transport network resources under OpenStack and OpenFlow orchestration [J]. Optics Express, 2014, 22(3): 2595–2602.

Part II
Novel Data Center Network Structures

Chapter 3
HCN: A Server-Centric Network Topology for Data Centers

Abstract A fundamental goal of data-center networking is to efficiently interconnect a large number of servers with the low equipment cost. Several server-centric network topologies for data centers have been proposed. They, however, are not truly expandable and suffer a low degree of regularity and symmetry. Inspired by the commodity servers in today's data centers that come with dual-port, we consider how to build expandable and cost-effective topologies without expensive high-end switches and additional hardware on servers except the two NIC ports. In this chapter, two such network topologies, called HCN and BCN, are designed based on the compound graph theory. The two topologies can achieve hierarchical expansion by only adding a small number of links among servers. Although the server degree is only 2, HCN can be expanded very easily to encompass hundreds of thousands servers with the low diameter and high bisection width. HCN also offers high degree of regularity, scalability and symmetry, which conform to the modular design of data centers. BCN is the largest known network topology for data centers with the server degree 2 and network diameter 7. BCN has many attractive features, including the low diameter, high bisection width, large number of node-disjoint paths for the one- to-one traffic, and good fault-tolerant ability. Mathematical analysis and comprehensive simulations show that HCN and BCN possess excellent topological properties.

3.1 Introduction

A fundamental goal of data center networking (DCN) is to efficiently interconnect a large number of servers with the low equipment cost. Traditional data center networks mainly interconnect servers into tree-like topologies through switches, core switches and core routers, which propose high demand to upper-layer network devices, and make these network devices easily become the performance bottleneck. Besides, the expansion of data centers usually introduces expensive high-end core routers and core switches, leading to high expansion cost. Additionally, tree-like topologies has poor fault-tolerance, and is prone to single-point failure. These topologies are

© Springer Nature Singapore Pte Ltd. 2022
D. Guo, *Data Center Networking*,
https://doi.org/10.1007/978-981-16-9368-7_3

hard to meet the data center design goals, including the low equipment cost, high network capacity, support of incremental expansion, and robustness. A number of novel DCN network topologies are proposed recently and can be roughly divided into two categories. One is switch-centric, which organizes switches into multi-layer tree topologies. Fat-Tree [1], VL2 [2] fall into such a category. The other is server-centric. DCell [3], FiConn [4, 5], BCube [6] and MDCube [7] fall into the second category.

With the increasing multi NICs and data forwarding capabilities of servers, the efficient data center networks can be achieved by direct connections between a large number of servers and low-end switches. These server-centric topologies puts the interconnection intelligence on servers and uses switches only as cross-bars. Compound graph and hierarchical networks become the mainstream design methods of server-centric data center network. Among others, a server-centric topology has the following advantages. First, each server participates into the routing process, so that there is no core router and core switch in the network, and the scalability is greatly enhanced. Second, in current practice, servers are more programmable than switches, so the deployment of new DCN topology is more feasible. Besides, multiple NIC ports in servers can be used to improve the end-to-end throughput as well as the fault-tolerant ability.

DCell [3] and BCube [6] are the two most representative server-centric DCN, and their nice topological properties and efficient algorithms have been derived at the cost as follows. They use no less than 4 ports per server, and large number of switches and links, so as to scale to a large server population. If they use servers with only 2 ports, the server population is very limited and cannot be enlarged since they are at most two levels. When network topologies are expanded to one higher level, DCell and BCube add one NIC and link for each existing server, and BCube has to be appended large number of additional switches. Thus the time and human power needed to upgrade tens or hundreds of thousands servers are very expensive.

Such topologies, however, are not truly expandable. A network is expandable if no changes with respect to the node's configuration and link connections are necessary when it is expanded. This may cause negative influence on applications running on all of the existing servers during the process of topology expansion. In this chapter, we focus on the interconnection of a large number of commodity dual-port servers since such servers are already available in current practice. It is challenging to interconnect a large population of such servers in data centers, because we should also guarantee the low diameter and the high bisection width.

In this chapter, we first propose a hierarchical irregular compound network, denoted as HCN, which can be expanded by only adding one link to a few number of servers. Moreover, HCN offers a high degree of regularity, scalability and symmetry, which conform to the modular designs of data centers. Inspired by the smaller network size of HCN compared to FiConn, we further study the degree/diameter problem [8, 9] in the scenario of building a scalable network topology for data centers using dual-port servers. Given the maximum node degree and network diameter, the degree/diameter problem aims to determine the largest graphs. Although many efforts [10, 11] have been made to study the degree/diameter problem in graph theory, it is still open in the field of DCN.

We then propose BCN, a Bidimensional Compound Network for data centers, which inherits the advantages of HCN. BCN is a level-i irregular compound graph recursively defined in the first dimension for $i \geq 0$, and a level one regular compound graph in the second dimension. In each dimension, a high-level BCN employs a one lower level BCN as a unit cluster and connects many such clusters by means of a complete graph. BCN of level one in each dimension is the largest known network topology for data centers, with the server degree 2 and the network diameter 7. In this case, the order of BCN is significantly larger than that of FiConn(n, 2), irrespective of the value of n. Besides such advantages, BCN has other attractive properties, including the low diameter and cost, high bisection width, high path diversity for the one-to-one traffic, good fault-tolerant ability, and relative shorter fault-tolerant path than FiConn.

3.2 The Design of HCN Topology

Hierarchical network is a natural way to construct large networks. In a hierarchical network, lower level networks support local communications, while higher level networks support remote communications. The compound graph is suitable for designing large-scale networks, due to its good regularity and expandability, and has been widely used in practice. We first review the definitions of regular graph and compound graph.

3.2.1 Basic Theories of the Compound Graph

Definition 3.1 A regular graph is a graph where each vertex has the same degree. A regular graph with vertices of degree k is called a k-regular graph.

Definition 3.2 Given two regular graphs G and G_1, a level-1 regular compound graph $G(G_1)$ is obtained by replacing each node of G by a copy of G_1 and replacing each link of G by a link that connects two corresponding copies of G_1.

As shown in Fig. 3.1, a level-1 regular compound graph $G(G_1)$ employs G_1 as a unit cluster and connects many such clusters by means of a regular graph G. In the resultant graph, the topology of G is preserved, and only one link is inserted to connect two copies of G_1. An additional remote link is associated with each node in a cluster. A constraint must be satisfied for the two graphs to constitute a regular compound graph. The node degree of G must be equal to the number of nodes in G_1. Otherwise, an irregular compound graph is obtained.

The basic idea of a compound graph can be extended to the context of a multi-level compound graph, recursively. For ease of explanation, we consider the case where the

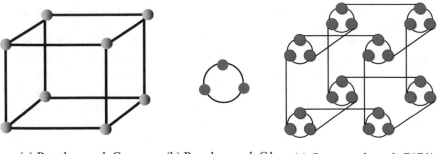

(a) Regular graph G   (b) Regular graph G1   (c) Compound graph G(G1)

Fig. 3.1 An illustrative example of the compound graph, which interconnects eight rings by means of the three-dimensional hypercube

Table 3.1 Summary of main notations

Term	Definition
α	Number of master servers in the level-0 BCN
β	Number of slave servers in the level-0 BCN
n	$n = \alpha + \beta$ is the number of ports of a mini-switch
h	Level of BCN in the first dimension
γ	Level of the unit BCN in the second dimension
$s_h = \alpha^h \beta$	Number of slave servers in any give BCN(α, β, h)
$s_\gamma = \alpha^\gamma \beta$	Number of slave servers in BCN(α, β, γ)
BCN$(\alpha, \beta, 0)$	Level-0 BCN, i.e., the smallest building block
BCN(α, β, h)	Level-h BCN in the first dimension
$G(BCN(\alpha, \beta, h))$	A compound graph that uses BCN(α, β, h) as G_1 and a complete graph as G
BCN$(\alpha, \beta, h, \gamma)$	A general BCN that always expands in the first dimension while only expands in the second dimension when $h \geq \gamma$
u	Order of BCN(α, β, h) in BCN$(\alpha, \beta, h, \gamma)$ from the viewpoint of the second dimension
v	Order of a $G(BCN(\alpha, \beta, \gamma))$ in BCN$(\alpha, \beta, h, \gamma)$ from the viewpoint of the first dimension

regular G is a complete graph. A level-2 compound graph $G^2(G_1)$ employs $G(G_1)$ as a unit cluster and connects many such clusters using a complete graph G. More generically, a level-i ($i > 0$) graph $G^i(G_1)$ adopts a level-$(i - 1)$ graph $G^{i-1}(G_1)$ as a unit cluster and connects many such clusters by a complete graph G.

Among existing DCNs, DCell and FiConn are constructed recursively by compound graph. In each iteration process, DCell forms a regular compound graph, while FiConn forms a irregular compound graph. HCN and BCN are introduced in this chapter. They are constructed by means of compound graph. Table 3.1 lists the notations used in the rest of this chapter.

3.2.2 The Construction Methodology of HCN

HCN is constructed based on irregular compound graph. It is defined by a recursive way as follows. For any given $h \geq 0$, we denote a level-h irregular compound network as HCN(n, h). HCN is a recursively defined topology. A high-level HCN(n, h) employs a low level HCN$(n, h - 1)$ as a unit cluster and connects many such clusters by means of a complete graph. HCN$(n, 0)$ is the smallest module (basic construction unit) that consists of n dual-port servers and a n-port mini-switch. For each server, its first port is used to connect with the mini-switch while the second port is employed to interconnect with another server in different smallest modules for constituting larger networks. A server is *available* if its second port has not been connected.

HCN$(n, 1)$ is constructed using n basic modules HCN$(n, 0)$. In HCN$(n, 1)$, there is only one link between any two basic modules by connecting two available servers that belong to different basic modules. Consequently, for each HCN$(n, 0)$ inside HCN$(n, 1)$ all of the servers are associated with a level-1 link except one server that is reserved for the construction of HCN$(n, 2)$. Thus, there are n available servers in HCN$(n, 1)$ for further expansion at a higher level. Similarly, HCN$(n, 2)$ is formed by n level-1 HCN$(n, 1)$s, and has n available servers for interconnection at a higher level.

In general, HCN(n, i) for $i \geq 0$ is formed by n HCN$(n, i - 1)$s, and has n available servers each in one HCN$(n, i - 1)$ for further expansion. According to Definition 3.2, HCN(n, i) acts as G_1 and a complete graph of n nodes acts as G. Here, $G(G_1)$ produces an irregular compound graph since the number of available servers in HCN(n, i) is n while the node degree of G is $n - 1$. To facilitate the construction of any level-h HCN, we introduce Definition 3.3 as follows.

Definition 3.3 Each server in HCN(n, h) is assigned a label $x_h \ldots x_1 x_0$, where $1 \leq x_i \leq n$ for $0 \leq i \leq h$. Two servers $x_h \ldots x_1 x_0$ and $x_h \ldots x_{j+1} x_{j-1} x_j^j$ are connected only if $x_j \neq x_{j-1}, x_{j-1} = x_{j-2} = \cdots = x_1 = x_0$ for some $1 \leq j \leq h$, where $1 \leq x_0 \leq \alpha$ and x_j^j represents j consecutive x_js. Here, n servers are reserved for further expansion only if $x_h = x_{h-1} = \cdots = x_0$ for any $1 \leq x_0 \leq n$.

Figure 3.2 plots an example of HCN$(4, 2)$ constructed according to Definition 3.3. HCN$(4, 2)$ consists of four HCN$(4, 1)$s and a HCN $(4, 1)$ has four HCN$(4, 0)$s. The second port of four servers, 111, 222, 333, and 444, are reserved for further expansion.

In a level-h HCN, each server recursively belongs to level-0, level-1, level-2, ..., level-h HCNs, respectively. Similarly, any lower level HCN belongs to many higher level HCNs. To characterize such a property, let x_i indicate the order of HCN$(n, i - 1)$, containing a server $x_h \ldots x_1 x_0$, among all of the level-$(i - 1)$ HCNs of HCN(n, i) for $1 \leq i \leq h$. We further use $x_h x_{h-1} \ldots x_i (1 \leq i \leq h)$ as a prefix to indicate HCN$(n, i - 1)$ that contains such a server in HCN(n, h). We use the server 423 as an example. Here, $x_1 = 2$ indicates the 2th HCN$(4, 0)$ in HCN$(4, 1)$ that contains such a server. Such a HCN$(4, 0)$ contains the servers 421, 422, 423, and 444. Here, $x_2 = 4$ indicates the 4th level-1 HCN in a level-2 HCN that contains such

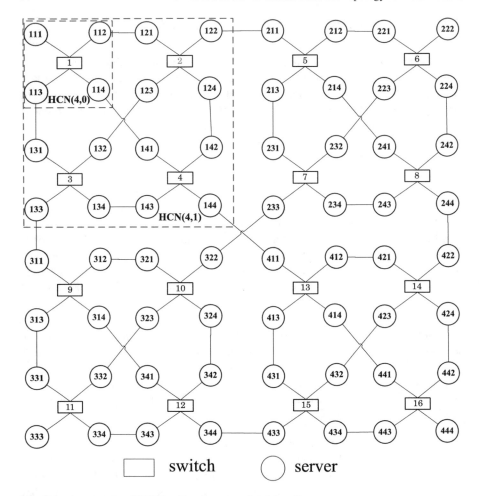

Fig. 3.2   An example of HCN(n, h), where $n = 4$ and $h = 2$

a server. Thus, $x_2 x_1 = 42$ indicates the level-0 HCN that contains the server 423 in a level-2 HCN.

HCN offers a high degree of regularity, scalability and symmetry. Additionally, HCN owns two topological advantages, i.e., expandability and equal server degree. Compared with FiConn, HCN has the benefits of easy implementation and low cost. Its network order, however, is less than that of FiConn in the same setting and we thus study the degree/diameter problem of DCN.

3.3 The Design of BCN Topology

3.3.1 Description of BCN Topology

BCN is a multi-level irregular compound graph recursively defined in the first dimension, and a level one regular compound graph in the second dimension. In each dimension, a high-level BCN employs a one low level BCN as a unit cluster and connects many such clusters by means of a complete graph.

Let BCN($\alpha, \beta, 0$) denote the basic building block, where $\alpha + \beta = n$. It has n servers and one n-port mini-switch. All of the servers are connected to the mini-switch using their first ports, and are partitioned into two disjoint groups, referred to as the *master* and *slave* servers. Let α and β be the number of master servers and slave servers, respectively. Here, servers really do not have master/slave relationship in functionality. The motivation of such a partition is just to ease the presentation. As discussed later, the second port of master servers and slave servers are used to constitute larger BCNs in the first and second dimensions, respectively. A server is available indicates that the second ports of the server are still idle.

1. Hierarchical BCN in the first dimension

For any given $h \geq 0$, we use BCN(α, β, h) to denote a level-h BCN formed by all of the *master* servers in the first dimension. For any $h > 1$, BCN(α, β, h) is an irregular compound graph, where G is a complete graph with α nodes while G_1 is BCN($\alpha, \beta, h - 1$) with α available master servers. It is worth noticing that, for any $h \geq 0$, BCN(α, β, h) still has α available master servers for further expansion, and is equivalent to HCN(α, h). The only difference is that each mini-switch also connects β slave servers besides α master servers in BCN(α, β, h).

2. Hierarchical BCN in the second dimension

There are β available slave servers in the smallest module BCN($\alpha, \beta, 0$). In general, there are $s_h = \alpha^h \cdot \beta$ available slave servers in any given BCN(α, β, h) for $h \geq 0$. We study how to utilize those available slave servers to expand BCN(α, β, h) from the second dimension. A level-1 regular compound graph, $G(BCN(\alpha, \beta, h))$, is a natural way to realize such a goal. It uses BCN(α, β, h) as a unit cluster and connects $s_h + 1$ copies of BCN(α, β, h) by means of a complete graph using the second ports of all of available slave servers. The resultant $G(BCN(\alpha, \beta, h))$ cannot be further expanded in the second dimension since it has no available slave servers. It, however, still can be expanded in the first dimension without destroying the existing network. To ease further discussion, we first give Theorem 3.1 and its proof.

Theorem 3.1 *The total number of slave servers in any given BCN(α, β, h) is*

$$s_h = \alpha^h \cdot \beta. \tag{3.1}$$

Fig. 3.3 A
$G(BCN(4, 4, 0))$ topology
that consists of slave servers
in five BCN(4, 4, 0)s in the
second dimension

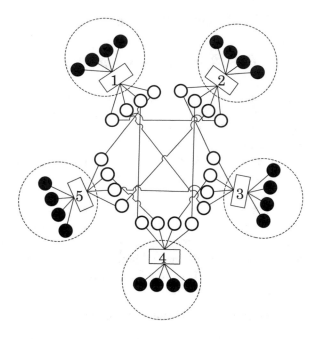

Proof We know that any given BCN(α, β, i) is built with α copies of a lower-level BCN($\alpha, \beta, i - 1$) for $1 \leq i$. Thus, it is reasonable that BCN(α, β, h) has α^h smallest module BCN($\alpha, \beta, 0$)s. In addition, each smallest module has β slave servers. Consequently, the total number of slave servers in BCN(α, β, h) is $s_h = \beta \cdot \alpha^h$. Thus proved. $\square$

Figure 3.3 plots an example of $G(BCN(4, 4, 0))$. The four slave servers connected with a mini-switch in BCN(4, 4, 0) is the unit cluster. A complete graph is used to connect five copies of BCN(4, 4, 0). Consequently, only one remote link is associated with each slave server in a unit cluster. Thus, the node degree is two for each slave server in the resultant network.

3. Bidimensional hierarchical BCN

After separately designing BCN(α, β, h) and $G(BCN(\alpha, \beta, h))$, we design a scalable bidimensional BCN formed by both master and slave servers. Let BCN(α, β, h, γ) denote a bidimensional BCN, where h denotes the level of BCN in the first dimension, and γ denotes the level of BCN that is selected as the unit cluster in the second dimension. In this case, BCN($\alpha, \beta, 0$) consists of α master servers, β slave servers and one n-port switch, i.e., it is still the smallest module of any level bidimensional BCN.

To increase servers in data centers on-demand, it is required to expand an initial lower-level BCN(α, β, h) from the first or second dimension without destroying the existing topology. A bidimensional BCN is always BCN(α, β, h) as h increases when $h < \gamma$. In such a scenario, the unit cluster for expansion in the second dimension has not been formed. When h increases to γ, we achieve BCN(α, β, γ) in the first dimension, and it can be used as the unit cluster for expansion in the second dimension. In the resultant BCN($\alpha, \beta, \gamma, \gamma$), there are $\alpha^\gamma \cdot \beta + 1$ copies of BCN(α, β, γ) and α available master servers in each BCN(α, β, γ). A sequential number u is employed to identify BCN(α, β, γ) among $\alpha^\gamma \cdot \beta + 1$ ones in the second dimension, where u ranges from 1 to $\alpha^\gamma \cdot \beta + 1$. Fig. 3.3 plots an example of BCN(4, 4, 0, 0) consisting of five BCN(4, 4, 0)s, where $h = r = 0$. It is worth noticing that BCN($\alpha, \beta, \gamma, \gamma$) cannot be further expanded in the second dimension since it has no available slave servers. It, however, still can be expanded in the first dimension without destroying the existing network in the following way.

When $h > \gamma$, each BCN(α, β, γ) in BCN($\alpha, \beta, \gamma, \gamma$) becomes BCN($\alpha, \beta, h$) in the first dimension. There are $\alpha^{h-\gamma}$ homogeneous BCN(α, β, γ)s inside each BCN(α, β, h). Thus, we use a sequential number v to identify BCN(α, β, γ) in each BCN(α, β, h) in the first dimension, where v ranges from 1 to $\alpha^{h-\gamma}$. Thus, the coordinate of each BCN(α, β, γ) in the resultant topology is denoted by a pair of v and u. It is worth noticing that only those BCN(α, β, γ)s with $v = 1$ in the resultant topology are connected by a complete graph in the second dimension, and form the first $G(BCN(\alpha, \beta, \gamma))$. Consequently, messages between any two servers in different BCN(α, β, γ)s with the same value of v except $v = 1$ must be relayed by related BCN(α, β, γ) in the first $G(BCN(\alpha, \beta, \gamma))$. Thus, the first $G(BCN(\alpha, \beta, \gamma))$ becomes a bottleneck of the resultant topology. To address such an issue, all of BCN(α, β, γ)s with $v = i$ are also connected by means of a completed graph so as to produce the ith $G(BCN(\alpha, \beta, \gamma))$, for other values of v besides 1. By now, we achieve BCN(α, β, h, γ) in which each $G(BCN(\alpha, \beta, \gamma))$ is a regular compound graph, where G is a complete graph with $\alpha^\gamma \cdot \beta$ nodes and G_1 is BCN(α, β, γ) with $\alpha^\gamma \cdot \beta$ available slave servers.

Figure 3.4 plots BCN(4, 4, 1, 0) formed by all of the master and slave servers from the first and second dimensions. Note that only the first and third BCN(4, 4, 1)s are plotted, while other three BCN(4, 4, 1)s are not shown due to page limitations. We can see that BCN(4, 4, 1, 0) has five homogeneous BCN(4, 4, 1)s in the second dimension and four homogeneous $G(BCN(4, 4, 0))$s in the first dimension. In the resultant topology, the node degree of each slave server is two while that of each master server is at least one and at most two.

3.3.2 The Construction Methodology of BCN

A higher level BCN network can be built by an incremental expansion using one lower level BCN as a unit cluster and connecting many such clusters by means of a complete graph. Specific construction methods are as follows.

1. When $h < \gamma$: BCN(α, β, h) can be achieved by the construction methodology of HCN(α, h).

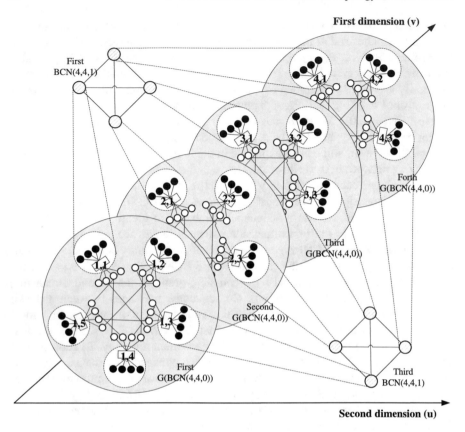

Fig. 3.4 An illustrative example of BCN(4, 4, 1, 0)

2. When $h = \gamma$: All of the slave servers in BCN(α, β, γ) are utilized for expansion
   in the second dimension. Each slave server in BCN(α, β, γ) is identified by a
   unique label $x = x_\gamma \ldots x_1 x_0$ where $1 \leq x_i \leq \alpha$ for $1 \leq i \leq \gamma$ and $\alpha + 1 \leq x_0 \leq$
   n. Besides the unique label, each slave server can be equivalently identified by a
   unique $id(x)$ that denotes its order among all of the slave servers in BCN(α, β, γ)
   and ranges from 1 to s_γ. For each slave server, the mapping between a unique
   $id(x)$ and its label is bijection, as defined in Theorem 3.2.

Theorem 3.2 *For any slave server* $x = x_\gamma \ldots x_1 x_0$, *its unique* $id(x)$ *is given by*

$$id(x_\gamma \ldots x_1 x_0) = \sum_{i=1}^{\gamma} (x_i - 1) \cdot \alpha^{i-1} \cdot \beta + (x_0 - \alpha). \tag{3.2}$$

Proof x_i denotes the order of BCN($\alpha, \beta, i - 1, \gamma$) that contains the slave server x in
a higher level BCN(α, β, i, γ) for $1 \leq i \leq \gamma$. In addition, the total number of slave
servers in any BCN($\alpha, \beta, i - 1, \gamma$) is $\alpha^{i-1} \cdot \beta$. Thus, there exist $\sum_{i=1}^{\gamma} (x_i - 1) \cdot \alpha^{i-1} \cdot$

β slave servers in other smallest modules before the smallest module BCN($\alpha, \beta, 0$) that contains the server x. On the other hand, there are other $x_0 - \alpha$ slave servers that reside in the same smallest module with the server x but has a lower x_0 than the server x. Thus proved. □

As mentioned above, the resultant BCN network when $h = \gamma$ is a $G(BCN(\alpha, \beta, \gamma))$ consisting of $s_\gamma + 1$ copies of a unit cluster BCN(α, β, γ). In such a case, BCN$_u(\alpha, \beta, \gamma)$ denotes the uth unit cluster in the second dimension. In BCN$_u(\alpha, \beta, \gamma)$, each server is assigned a unique label $x = x_\gamma \ldots x_1 x_0$ and a 3-tuple $[v(x) = 1, u, x]$, where $v(x)$ is defined in Theorem 3.3. In BCN$_u(\alpha, \beta, \gamma)$, all of the master servers are interconnected according to the rules in Definition 3.3 for $1 \le u \le s_\gamma + 1$.

Many different ways can be used to interconnect all of the slave servers in $s_\gamma + 1$ homogeneous BCN(α, β, γ)s to constitute a $G(BCN(\alpha, \beta, \gamma))$. For any two slave servers $[1, u_s, x_s]$ and $[1, u_d, x_d]$, as mentioned in literature [12, 13] they are interconnected only if

$$\left. \begin{array}{l} u_d = (u_s + id(x_s)) \bmod (s_\gamma + 2) \\ id(x_d) = s_\gamma + 1 - id(x_s) \end{array} \right\} \qquad (3.3)$$

where $id(x_s)$ and $id(x_d)$ are calculated by Formula 3.2. In literature [7], the two slave servers are connected only if

$$\left. \begin{array}{l} u_s > id(x_s) \\ u_d = id(x_s) \\ id(x_d) = (u_s - 1) \bmod s_\gamma \end{array} \right\} \qquad (3.4)$$

This chapter does not focus on designing new interconnection methods for all of the slave servers since the above two and other permutation methods are suitable to constitute $G(BCN(\alpha, \beta, \gamma))$.

(3) When $h > \gamma$: After achieving BCN($\alpha, \beta, \gamma, \gamma$), the resultant network can be incrementally expanded in the first dimension without destroying the existing topology. In this scenario, BCN(α, β, h, γ) consists of $s_\gamma + 1$ copies of a unit cluster BCN(α, β, h) in the second dimension. Each server in BCN$_u(\alpha, \beta, h)$ is assigned a unique label $x = x_h \ldots x_1 x_0$ for $1 \le u \le s_\gamma + 1$. In addition, BCN$_u(\alpha, \beta, h)$ has $\alpha^{h-\gamma}$ BCN(α, β, γ)s in the first dimension. Recall that a sequential number v is employed to rank those BCN(α, β, γ)s in BCN$_u(\alpha, \beta, h)$.

In BCN$_u(\alpha, \beta, \gamma)$, each server is assigned a 3-tuple $[v(x), u, x]$, where $v(x)$ is defined in Theorem 3.3. A pair of u and $v(x)$ is sufficient to identify the unit cluster BCN(α, β, γ) that contains the server x in BCN(α, β, h, γ). For a slave server x, we further assign a unique $id(x_\gamma \ldots x_1 x_0)$ to indicate the order of x among all of the slave servers in the same BCN(α, β, γ).

Theorem 3.3 *For any server labeled $x = x_h \ldots x_1 x_0$ for $h \ge \gamma$, the rank of the module BCN(α, β, γ) in BCN(α, β, h) the server x resides in is given by*

Algorithm 3.1 Construction of BCN(α, β, h, γ)

Require: $h > \gamma$

1: Connects all of the servers that have the same u and the common length-h prefix of their labels to the same min-switch using their first ports.   ▷ Construction of all smallest modules BCN($\alpha, \beta, 0$)

2: **for** [**do**Interconnect master servers that hold the same u to form $\alpha^\gamma \cdot \beta + 1$ copies of BCN(α, β, h)] $u = 1$ to $\alpha^\gamma \cdot \beta + 1$

3:    Any master server $[v(x), u, x = x_h \ldots x_1 x_0]$ is interconnected with a master server $[v(x'), u, x' = x_h \ldots x_{j+1} x_{j-1} x_j^j]$ using their second ports if $x_j \neq x_{j-1}, x_{j-1} = \cdots = x_1 = x_0$ for some $1 \leq j \leq h$, where $1 \leq x_0 \leq \alpha$ and a_j^j represents j consecutive a_js.

4: **for** [**do**Connect slave servers that hold the same v to form the vth G(BCN(α, β, γ)) in BCN(α, β, h, γ)]$v = 1$ to $\alpha^{h-\gamma}$

5:    Interconnect any two slave servers $[v(x), u_x, x = x_h \ldots x_1 x_0]$ and $[v(y), u_y, y = y_h \ldots y_1 y_0]$ using their second ports only if (1) $v(x) = v(y)$; (2) $[u_x, x_\gamma \ldots x_1 x_0]$ and $[u_y, y_\gamma \ldots y_1 y_0]$ satisfy the constraints in Formula 3.3.

$$v(x) = \begin{cases} 1, & \text{if } h = \gamma \\ x_{\gamma+1}, & \text{if } h = \gamma + 1 \\ \sum_{i=\gamma+2}^{h}(x_i - 1) \cdot \alpha^{i-\gamma-1} + x_{\gamma+1}, & \text{if } h > \gamma + 1 \end{cases} \tag{3.5}$$

Proof Recall that any BCN(α, β, i) is constructed with α copies of BCN($\alpha, \beta, i - 1$) for $1 \leq i$. Therefore, the total number of BCN(α, β, γ)s in BCN(α, β, i) for $i > \gamma$ is $\alpha^{i-\gamma}$. In addition, x_i indicates BCN($\alpha, \beta, i - 1$) in the next higher level BCN(α, β, i) that contains such a server for $1 \leq i$. Thus, there are $(x_i - 1) \cdot \alpha^{i-\gamma-1}$ BCN(α, β, γ)s in other $x_i - 1$ previous BCN($\alpha, \beta, i - 1$)s inside BCN(α, β, i) for $\gamma + 2 \leq i \leq h$. In addition, $x_{\gamma+1}$ indicates the sequence of BCN(α, β, γ) in BCN($\alpha, \beta, \gamma + 1$) the server x resides in. Therefore, the rank of BCN(α, β, γ) in BCN(α, β, h) such a server resides in is given by Formula 3.5. Thus proved.   □

After assigning a 3-tuple to all of the master and slave servers, we propose a general procedure to constitute BCN(α, β, h, γ) ($h > \gamma$), as shown in Algorithm 3.1. The entire procedure includes three parts. The first part groups all of the servers into the smallest modules BCN($\alpha, \beta, 0$) for further expansion. The second part constructs $s_\gamma + 1$ homogeneous BCN(α, β, h)s by connecting the second ports of those master servers that have the same u and satisfy the constraints mentioned in Definition 3.3. Furthermore, the third part connects the second ports of those slave servers that have the same v and satisfy the constraints defined by Formula 3.3. Consequently, the construction produce results in BCN(α, β, h, γ) consisting of $\alpha^{h-\gamma}$ homogeneous G(BCN(α, β, γ))s. Note that it is not necessary that the connection rule of all of the slave servers must be that given by Formula 3.3. It also can be that defined by Formula 3.4.

3.4 Routing Mechanism of BCN

The unicast traffic is the basic traffic model and the good one-to-one support also results in the good many-to-one and many-to-many support. In this section, we start with the single-path routing for the unicast traffic in BCN without failures of switches, servers, and links. On this basis, we propose fault-tolerant routing schemes to address those representative failures by employing the benefits of multi-paths between any two servers.

3.4.1 Single-path for the Unicast Traffic

1. In the case of $h < \gamma$

For any $BCN(\alpha, \beta, h)$ $(1 \leq h)$, we propose an efficient routing scheme, denoted as *FdimRouting*, to find a single-path between any pair of servers in a distributed manner. Let src and dst denote the source and destination servers in the same $BCN(\alpha, \beta, h)$ but different $BCN(\alpha, \beta, h-1)$s. The source and destination can be of the master server or the slave server. The routing scheme first determines the link $(dst1, src1)$ that interconnects the two $BCN(\alpha, \beta, h-1)$s that src and dst are located at. It then derives two sub-paths from src to $dst1$ and from $src1$ to dst. The path from src to dst is the combination of the two sub-paths and the link $(dst1, src1)$. Each of the two sub-paths can be obtained by recursively invoking Algorithm 3.2.

In Algorithm 3.2, the labels of a pair of servers are retrieved from the two inputs that can be of 1-tuple or 3-tuple. A 3-tuple indicates that such a $BCN(\alpha, \beta, h)$ is a component of the entire $BCN(\alpha, \beta, h, \gamma)$ when $h \geq \gamma$. The *CommPrefix* calculates the common prefix of src and dst and the *GetIntraLink* identifies the link that connects the two sub-BCNs in $BCN(\alpha, \beta, h)$. Note that the two ends of the link can be directly derived from the indices of the two sub-BCNs according to Definition 3.3. Thus, the time complexity of *GetIntraLink* is $O(1)$.

In *FdimRouting* algorithm, the length of the path between two servers connecting to the same switch is one. Such an assumption was widely used in the designs of server-centric network topologies for data centers, such as DCell, FiConn, BCube, and MDCube. From the *FdimRouting*, we obtain the following Theorem 3.4.

Theorem 3.4 *The shortest path length among all of the server pairs in $BCN(\alpha, \beta, h)$ is at most $2^{h+1} - 1$.*

Proof For any two servers, src and dst, in $BCN(\alpha, \beta, h)$ but in different $BCN(\alpha, \beta, h-1)$s, let D_h denote the length of the single path resulted from Algorithm 3.2. The entire path consists of two sub-paths in two different $BCN(\alpha, \beta, h-1)$s and one link connects the two lower BCNs. It is reasonable to infer that $D_h = 2 \cdot D_{h-1} + 1$ for $h > 0$ and $D_0 = 1$. We can derive that $D_h = \sum_{i=0}^{h} 2^i$. Thus proved. □

Algorithm 3.2 FdimRouting(src, dst)

Require: src and dst are two servers in BCN(α, β, h) ($h < \gamma$).
1: The labels of the two servers are retrieved from the inputs, and are $src = s_h s_{h-1} \ldots s_1 s_0$ and
   $dst = d_h d_{h-1} \ldots d_1 d_0$, respectively.
2: $pref \leftarrow CommPrefix(src, dst)$
3: Let m denote the length of $pref$
4: **if** $m==h$ **then**
5:     Return (src, dst)   ▷ The servers connect to the same switch.
6: ($dst1, src1$)←$GetIntraLink(pref, s_{h-m}, d_{h-m})$
7: $head \leftarrow FdimRouting(src, dst1)$
8: $tail \leftarrow FdimRouting(src1, dst)$
9: Return $head + (dst1, src1) + tail$

GetIntraLink($pref, s, d$)

1: Let m denote the length of $pref$
2: $dst1 \leftarrow pref + s + d^{h-m}$ ▷ d^{h-m} represents $h - m$ consecutive d
3: $src1 \leftarrow pref + d + s^{h-m}$ ▷ s^{h-m} represents $h - m$ consecutive s
4: Return ($dst1, src1$)

Algorithm 3.3 BdimRouting(src, dst)

Require: src and dst are   denoted   as $[v(s_h \ldots s_1 s_0), u_s, s_h \ldots s_1 s_0]$ and
   $[v(d_h \ldots d_1 d_0), u_d, d_h \ldots d_1 d_0]$ in BCN($\alpha, \beta, h \geq \gamma, \gamma$).
1: **if** [**then**In the same BCN(α, β, h)]$u_s==u_d$
2:     Return $FdimRouting(src, dst)$
3: $v_c \leftarrow v(s_h \ldots s_1 s_0)$ ▷ v_c can also be $v(d_h \ldots d_1 d_0)$
4: ($dst1, src1$)←$GetInterLink(u_s, u_d, v_c)$
5: $head \leftarrow FdimRouting(src, dst1)$ ▷ Find a path from src to $dst1$ in the u_s^{th} BCN(α, β, h) of
   BCN(α, β, h, γ)
6: $tail \leftarrow FdimRouting(src1, dst)$ ▷ Find a path from $src1$ to dst in the u_d^{th} BCN(α, β, h) of
   BCN(α, β, h, γ)
7: Return $head + (dst1, src1) + tail$

GetInterLink(s, d, v)

1: Infer two slave servers $[s, x = x_h \ldots x_1 x_0]$ and $[d, y = y_h \ldots y_1 y_0]$ from the sth and d^{th}
   BCN(α, β, h) in BCN(α, β, h, γ) such that (1) $v(x) = v(y) = v$; (2) $[s, x_\gamma \ldots x_1 x_0]$ and
   $[d, y_\gamma \ldots x_1 x_0]$ satisfy the constraints defined by Formula 3.3.
2: Return ($[s, x], [d, y]$)

The time complexity of Algorithm 3.2 is $O(2^h)$ for deriving the entire path, and
can be reduced to $O(h)$ for deriving only the next hop.

2. In the case of $h \geq \gamma$

Consider the routing scheme in any BCN(α, β, h, γ) consisting of $\alpha^\gamma \cdot \beta + 1$ copies
of BCN(α, β, h) for $h \geq \gamma$. The *FdimRouting* scheme can discover a path only if
the two servers are located at the same BCN(α, β, γ). In other cases, *FdimRouting*
alone cannot guarantee to find a path between any pair of servers. To handle such
an issue, we propose the *BdimRouting* scheme for the cases that $h \geq \gamma$, as shown in
Algorithm 3.3.

For any two servers, src and dst, in $BCN(\alpha, \beta, h, \gamma)$ $(h \geq \gamma)$, Algorithm 3.3 invokes Algorithm 3.2 to discover the path between the two servers only if they are in the same $BCN(\alpha, \beta, h)$. Otherwise, it first identifies the link $(dst1, src1)$ that interconnects the $v(src)^{th}$ $BCN(\alpha, \beta, \gamma)$s of $BCN_{u_s}(\alpha, \beta, h)$ and $BCN_{u_d}(\alpha, \beta, h)$. Note that the link that connects the $v(dst)^{th}$ instead of the $v(src)^{th}$ $BCN(\alpha, \beta, \gamma)$s of $BCN_{u_s}(\alpha, \beta, h)$ and $BCN_{u_d}(\alpha, \beta, h)$ is an alternative link. Algorithm 3.2 then derives a sub-path from src to $dst1$ that are in the $v(src)^{th}$ $BCN(\alpha, \beta, \gamma)$ inside $BCN_{u_s}(\alpha, \beta, h)$ and finds another sub-path from $src1$ to dst that are in $BCN_{u_d}(\alpha, \beta, h)$ by invoking Algorithm 3.2. Consequently, the path from src to dst is the combination of the two sub-paths and the link $(dst1, src1)$. From the $BdimRouting$, we obtain the following theorem.

Theorem 3.5 *The shortest path length among all of the server pairs in $BCN(\alpha, \beta, h, \gamma)$ $(h > \gamma)$ is at most $2^{h+1} + 2^{\gamma+1} - 1$.*

Proof In Algorithm 3.3, the entire routing path from src to dst might contain an inter-link between $dst1$ and $src1$, a first sub-path from src to $dst1$ and a second sub-path from $src1$ to dst. The length of the first sub-path is $2^{\gamma+1} - 1$ since the two end servers are in the same $BCN(\alpha, \beta, \gamma)$. Theorem 3.4 shows that the maximum path length of the second sub-path is $2^{h+1} - 1$. Consequently, the length of the entire path from src to dst is at most $2^{h+1} + 2^{\gamma+1} - 1$. Thus proved. □

It is worth noticing that the $GetInterLink$ can directly derive the end servers of the link only based on the three inputs and the constraints in Formula 3.3. Thus, the time complexity of the $GetInterLink$ is $O(1)$. The time complexity of Algorithm 3.3 is $O(2^k)$ for deriving the entire path, and can be reduced to $O(k)$ for deriving only the next hop.

3.4.2 Multi-paths for Unicast Traffic

Before the introduction of multi-paths for unicast traffic in BCN, we first give the concept of parallel paths. Two parallel paths between a source server src and a destination server dst exist if the intermediate servers on one path do not appear on the other. Compared with single-path routing, multi-paths routing has greater fault tolerance. In this subsection, the method is discussed of generating multiple parallel paths between any two servers.

Lemma 3.1 *There are $\alpha - 1$ parallel paths between any two servers, src and dst, in $BCN(\alpha, \beta, h)$ but not in the same $BCN(\alpha, \beta, 0)$.*

Proof We show the correctness of Lemma 3.1 by constructing such $\alpha - 1$ paths. The construction procedure is based on the single-path routing, FdimRouting, in the case of $h < \gamma$. We assume that $BCN(\alpha, \beta, i)$ is the lowest level BCN that contains the two servers src and dst. The FdimRouting determines the link $(dst1, src1)$ that interconnects the two $BCN(\alpha, \beta, i - 1)$s each contains one of the two servers, and then builds the first path passing that link. There are α one lower level $BCN(\alpha, \beta, i-1)$ in $BCN(\alpha, \beta, i)$ that contains the dst. The first path does not pass other intermediate $BCN(\alpha, \beta, i)$s, while each of other $\alpha - 2$ parallel paths must traverse one intermediate $BCN(\alpha, \beta, i-1)$. Then we describe the construction method of these $\alpha - 2$ parallel paths.

Let $x_h \ldots x_1 x_0$ and $y_h \ldots y_1 y_0$ denote the labels of $src1$ and $dst1$, respectively. Now we construct the other $\alpha - 2$ parallel paths from src to dst. First, a server labeled $z = z_h \ldots z_1 z_0$ is identified as a candidate server of $src1$ only if z_{i-1} is different from x_{i-1} and y_{i-1} while other parts of its label is the same as that of the label of $src1$. It is clear that there exist $\alpha - 2$ candidate servers of $src1$. Second, we find a parallel path from src to dst by building a sub-path from the source src to an intermediate server z and a sub-path from z to the destination dst. The two sub-paths can be produced by the FdimRouting. So far, all of the $\alpha - 1$ parallel paths between any two servers are constructed. Thus proved. □

Note that, the $\alpha - 1$ parallel paths between src and dst are built in a fully distributed manner only based on the labels of the src and dst without any overhead of control messages. We use Fig. 3.2 as an example to show the three parallel paths between any two servers. The first path from 111 to 144 is $111 \rightarrow 114 \rightarrow 141 \rightarrow 144$, which is built by Algorithm 3.2. Other two paths are $111 \rightarrow 113 \rightarrow 131 \rightarrow 134 \rightarrow 143 \rightarrow 144$ and $111 \rightarrow 112 \rightarrow 121 \rightarrow 124 \rightarrow 142 \rightarrow 144$. We can see that the three paths are node-disjointed and thus are parallel.

As for $BCN(\alpha, \beta, \gamma, \gamma)$ with $\alpha^{\gamma}\beta + 1$ copies of $BCN(\alpha, \beta, \gamma)$, if src and dst reside in the same $BCN(\alpha, \beta, \gamma)$, there are $\alpha - 1$ parallel paths between src and dst according to Lemma 3.1. Otherwise, we assume A and B denote two $BCN(\alpha, \beta, \gamma)$s in which src and dst reside, respectively. In such a case, there exist $\alpha^{\gamma}\beta$ parallel paths between A and B since $BCN(\alpha, \beta, \gamma, \gamma)$ connects $\alpha^{\gamma}\beta + 1$ copies of $BCN(\alpha, \beta, \gamma)$ by means of a complete graph. In addition, Lemma 3.1 shows that there are only $\alpha - 1$ parallel paths between any two servers in $BCN(\alpha, \beta, \gamma)$, such as A and B. Accordingly, it is easy to infer that Lemma 3.2 holds.

Lemma 3.2 *There are $\alpha - 1$ parallel paths between any two servers in $BCN(\alpha, \beta, \gamma, \gamma)$ but not in the same $BCN(\alpha, \beta, 0)$.*

In the case that $h > \gamma$, $BCN(\alpha, \beta, \gamma)$ is the unit cluster of $BCN(\alpha, \beta, h, \gamma)$. Assume src and dst are labelled as $[v(s_h \ldots s_1 s_0), u_s, s_h \ldots s_1 s_0]$ and $[v(d_h \ldots d_1 d_0), u_d, d_h \ldots d_1 d_0]$, and reside in two unit clusters with labels $< v(s_h \ldots s_1 s_0), u_s >$ and $< v(d_h \ldots d_1 d_0), u_d >$, respectively. According to Lemmas 3.1 and 3.2, there are $\alpha - 1$ parallel paths between src and dst if $u_s = u_d$ or $v(s_h \ldots s_1 s_0) = v(d_h \ldots d_1 d_0)$. In other cases, we select $BCN(\alpha, \beta, \gamma)$ with label $< v(s_h \ldots s_1 s_0), u_d >$ as a relay cluster. As aforementioned, there are $\alpha^{\gamma}\beta$ parallel paths between the unit clusters $< v(s_h \ldots s_1 s_0), u_s >$ and $< v(s_h \ldots s_1 s_0), u_d >$, while only $\alpha - 1$ parallel paths between $< v(s_h \ldots s_1 s_0), u_d >$ and $< v(d_h \ldots d_1 d_0), u_d >$. In addition, Lemma 3.1 shows that there are only $\alpha - 1$ parallel paths between any two servers in the same unit cluster. Accordingly, $\alpha - 1$ parallel paths exist between src and dst. Actually, the number of parallel paths between src and dst is also $\alpha - 1$

for another relay cluster $< v(d_h \ldots d_1 d_0), u_s >$. The two groups of parallel paths only intersect inside the unit clusters $< v(s_h \ldots s_1 s_0), u_s >$ and $< v(d_h \ldots d_1 d_0), u_d >$. So far, it is easy to derive Theorem 3.6.

Theorem 3.6 *No matter whether $h \leq \gamma$, there are $\alpha - 1$ parallel paths between any two servers in $BCN(\alpha, \beta, h, \gamma)$ but not in the same $BCN(\alpha, \beta, 0)$.*

Although BCN has the capability of providing multi-paths for the one-to-one traffic, the existing routing schemes, including the FdimRouting and the BdimRouting, only exploit one path. To enhance the transmission reliability for the one-to-one traffic, we adapt the routing path when the transmission meets failures of a link, a server, and a switch. It is worth noticing that those parallel paths between any pair of servers pass through the common switch that connects the destination server in the last step. This does not hurt the fault-tolerant ability of those parallel paths except the switch connecting the destination fails. In such a rare case, at most one reachable path exists between two servers.

3.4.3 Fault-Tolerant Routing in BCN

Before the discussion of fault-tolerant routing, we first give the definition of a failed link that can summarize three representative failures in data centers.

Definition 3.4 A link $(src1, dst1)$ is called failed only if the head $src1$ does not fail, however, cannot communicate with the tail $dst1$ no matter whether they are connected to the same switch or not. The failures of $dst1$, link, and the switch that connects $src1$ and $dst1$ can result in a failed link.

To improve the fault tolerance of FdimRouting and the BdimRouting, two fault-tolerant routing techniques are used, i.e., the *local-reroute* and *remote-reroute*. The local-reroute *adjusts a routing path that consists of local links* on the basis of the FdimRouting. On the contrary, the remote-reroute modifies those *remote links* in a path derived by BdimRouting. All of the links that interconnect master servers using the second ports are called the *local links*, while those links that interconnect slave servers using the second ports are called the *remote links*.

1. Local-reroute

Given any two servers src and dst in $BCN(\alpha, \beta, h, \gamma)$ $(h < \gamma)$, we can calculate a path from src to dst using the FdimRouting. Consider any failed link $(src1, dst1)$ in such a path, where $src1$ and $dst1$ are labeled $x_h \ldots x_1 x_0$ and $y_h \ldots y_1 y_0$, respectively. The FdimRouting does not take failed links into account. We introduce the *local-reroute* to bypass failed links by making local decisions. Here, each server has only local information. That is, it knows only the health state of the other servers on its directly connected switch (the master and slave servers) and the server connected over its second NIC (if any). Each server computes a set of relay servers for a failed next-hop server on demand. The assumption is that if the direct next-hop is not reachable at least one of the relay servers can be reachable from the current server.

The basic idea of the *local-reroute* is that $src1$ immediately identifies all of the usable candidate servers of $dst1$ and then selects one of such servers as a *relay* server. The server $src1$ first routes packets to *relay* along a path derived by the *FdimRouting* and then to the final destination dst along a path from *relay* to *dst*. If any link in the first sub-path from $src1$ to *relay* fails, the packets are routed towards dst along a new *relay* of the tail of the failed link, and then all of the existing *relay* servers in turn.

A precondition of the *local-reroute* is that $src1$ can identify a *relay* server for $dst1$ by only local decisions. Let m denote the length of the longest common prefix of $src1$ and $dst1$. Let $x_h \ldots x_{h-m+1}$ denote the longest common prefix of $src1$ and $dst1$ for $m \geq 1$. If $m \neq h$, the two servers $dst1$ and $src1$ are not connected with the same switch, and then the label $z_h \ldots z_1 z_0$ of the *relay* server can be given by

$$\left. \begin{array}{l} z_h \cdots z_{h-m+1} = y_h \cdots y_{h-m+1} \\ z_{h-m} \in \{\{1, 2, \ldots, \alpha\} - \{x_{h-m}, y_{h-m}\}\} \\ z_{h-m-1} \cdots z_1 z_0 = y_{h-m-1} \cdots y_1 y_0 \end{array} \right\} \tag{3.6}$$

Otherwise, we first derive the server $dst2$ that connects with the server $dst1$ using their second ports. The failure of the link $(src1, dst1)$ is equivalent to the failure of the link $(dst1, dst2)$ unless $dst1$ is the destination. Thus, we can derive a *relay* server of the server $dst2$ using Formula 3.6, i.e., the *relay* server of the server $dst1$. In summary, the total number of such relay servers is $\alpha - 2$, where $\alpha \approx (2 \cdot \gamma \cdot n)/(2 \cdot \gamma + 1)$ as shown in Theorem 3.8. It is unlikely that all of relay servers for a failed one-hop server will be unreachable simultaneously since the switch ports, n, in a data center is typical not small.

In Formula 3.6, the notation $h - m$ indicates that the two servers $src1$ and $dst1$ are in the same BCN$(\alpha, \beta, h - m)$ but in two different BCN$(\alpha, \beta, h - m - 1)$s. There exist α BCN$(\alpha, \beta, h - m - 1)$ subnets inside such a BCN$(\alpha, \beta, h - m)$. When $src1$ finds the failure of $dst1$, it chooses one *relay* server from all of BCN$(\alpha, \beta, h - m - 1)$ subnets in such a BCN$(\alpha, \beta, h - m)$ except the two subnets that contain $src1$ or $dst1$. If $src1$ selects a relay server for $dst1$ from BCN$(\alpha, \beta, h - m - 1)$ that contains $src1$, the packets will be routed back to $dst1$ that fails to route those packets.

For ease of understanding, we discuss an illustrative example of local-reroute. In Fig. 3.2, $111 \rightarrow 114 \rightarrow 141 \rightarrow 144 \rightarrow 411 \rightarrow 414 \rightarrow 441 \rightarrow 444$ is the path from 111 to 444 derived by Algorithm 3.2. Once the link $144 \rightarrow 411$ and/or the server 411 fails, the server 144 immediately finds server 211 or 311 as a relay server, and calculates a path from it to the relay server. If the relay server is 211, the path derived by Algorithm 3.2 is $144 \rightarrow 142 \rightarrow 124 \rightarrow 122 \rightarrow 122 \rightarrow 211$. After receiving packets towards 444, the derived path by Algorithm 3.2 from 211 to 444 is $211 \rightarrow 214 \rightarrow 241 \rightarrow 244 \rightarrow 422 \rightarrow 424 \rightarrow 442 \rightarrow 444$. It is worth noticing that if any link in the sub-path from 144 to 221 fails, the head of that link must bypass such a failed link and reaches 221 in the same way. If the link $122 \rightarrow 211$ fails, the server 311 will replace 211 as the relay server of 411. If there is a failed link in the sub-path from 211 to 444, the *local-reroute* is used to address the failed link in the same way.

It is worth noticing that the failed link (141, 144) will be found if the server 144 in the path from 111 to 444 fails. The failure of a link (141, 144) is equivalent to the failure of the link (144, 411). Hence, the servers 211 and 311 are the *relay* servers derived by the aforementioned rules and Formula 3.6. All of the servers each with an identifier starting with 1 from left to right cannot be the relay server since the path from the relay server to the destination will pass the failed link again.

If a server prefers to pre-compute and store the relay servers, it has to keep a forwarding table for its one-hop neighbors in the first dimension. Such forwarding table for relaying purpose is of size α, with each entry is of size $\alpha - 2$ since there exist $\alpha - 2$ relay servers for a failed one-hop server. Such a method incurs less delay than computing the relay servers on demand, however, consumes additional storage space of $O(\alpha^2)$. Such an overhead can be reduced to α if only a few relay servers are stored in each entry, e.g., three relay servers if such a number is enough for bypassing a failed one-hop server.

2. Remote-reroute

For any two servers, src and dst, in $BCN(\alpha, \beta, h, \gamma)$ $(h \geq \gamma)$, their 3-tuples are $[v_s, u_s, s_h \ldots s_1 s_0]$ and $[v_d, u_d, d_h \ldots d_1 d_0]$, respectively. The *local-reroute* can handle any failed link in the path from src to dst if they are in the same $BCN(\alpha, \beta, h)$ in $BCN(\alpha, \beta, h, \gamma)$, i.e., $u_s = u_d$. Otherwise, a pair of servers $dst1$ and $src1$ are derived according to the *GetInterLink* operation in Algorithm 3.3, and are denoted as $[u_s, v_s, x = x_h \ldots x_1 x_0]$ and $[u_d, v_s, y = y_h \ldots y_1 y_0]$, respectively. Clearly, $dst1$ and $src1$ are in the v_s^{th} $BCN(\alpha, \beta, \gamma)$s inside $BCN_{u_s}(\alpha, \beta, h)$ and $BCN_{u_d}(\alpha, \beta, h)$, respectively. The link $(dst1, src1)$ is the only one that interconnects the two v_s^{th} $BCN(\alpha, \beta, \gamma)$s in the two $BCN(\alpha, \beta, h)$s.

If the packets from src to dst meets failed links in the two sub-paths from src to $dst1$ and from $src1$ to dst, the *local-reroute* can address those failed links. The *local-reroute*, however, cannot handle the failures of $dst1$, $src1$, and the links between them. In such cases, the packets cannot be forwarded from the u_s^{th} $BCN(\alpha, \beta, h)$ to the u_d^{th} $BCN(\alpha, \beta, h)$ inside such a $BCN(\alpha, \beta, h, \gamma)$ through the desired link $(dst1, src1)$. We propose the *remote-reroute* to address such an issue.

The basic idea of *remote-reroute* is to transfer the packets to another slave server $dst2$ that is connected with the same switch together with $dst1$ if at least one such slave server and its associated links are usable. The label of $dst2$ is $x_h \ldots x_1 x_0'$ where x_0' can be any integer ranging from $\alpha + 1$ to n except x_0. Assume that the other end of the link that is incident from $dst2$ using its second port is a slave server $src2$ in another $BCN_{u_i}(\alpha, \beta, h)$ inside the entire network. The packets are then forwarded to the slave server $src2$, and are routed to the destination dst along a path derived by Algorithm 3.3. If a link in the path from $src2$ to dst fails, the *local-reroute*, *remote-reroute*, and Algorithm 3.3 can handle the failed links.

3.5 Performance Evaluation

In this section, we analyze several basic topological properties of HCN and BCN, including the network order, network diameter, server degree, connectivity, and path diversity. Then we conduct simulations to evaluate the distribution of path length, average path length, and the robustness of routing algorithms.

3.5.1 Network Order

Lemma 3.3 *The total number of servers in BCN(α, β, h) is $\alpha^h \cdot (\alpha + \beta)$, including α^{h+1} master and $\alpha^h \cdot \beta$ slave servers.*

Proof As mentioned above, any given level BCN consists of α one lower BCNs. There are α^h level-0 BCNs in BCN(α, β, h), where a level-0 BCN consists of α master servers and β slave servers. Thus proved.   □

Lemma 3.4 *The number of servers in $G(BCN(\alpha, \beta, h))$ is $\alpha^h \cdot (\alpha + \beta) \cdot (\alpha^h \cdot \beta + 1)$, including $\alpha^{h+1} \cdot (\alpha^h \cdot \beta + 1)$ and $\alpha^h \cdot \beta \cdot (\alpha^h \cdot \beta + 1)$ master and slave servers, respectively.*

Proof As mentioned above, there are $\alpha^h \cdot \beta + 1$ copies of BCN(α, β, h) in $G(BCN(\alpha, \beta, h))$. In addition, the number of servers in BCN(α, β, h) has been proved by Lemma 3.3. Thus proved.   □

Theorem 3.7 *The number of servers in BCN(α, β, h, γ) is*

$$
\begin{cases}
\alpha^h \cdot (\alpha + \beta), & \text{if } h < \gamma \\
\alpha^{h-\gamma} \cdot \left(\alpha^\gamma \cdot (\alpha + \beta) \cdot (\alpha^\gamma \cdot \beta + 1) \right), & \text{if } h \geq \gamma
\end{cases}
\tag{3.7}
$$

Proof Lemma 3.3 has proved such an issue when $h < r$. when $h = r$, BCN($\alpha, \beta, \gamma, \gamma$) is just $G(BCN(\alpha, \beta, \gamma))$. Thus, there are $\alpha^\gamma \cdot (\alpha + \beta) \cdot (\alpha^\gamma \cdot \beta + 1)$ servers in BCN($\alpha, \beta, \gamma, \gamma$) according to Lemma 3.4. In addition, BCN(α, β, h, γ) contains $\alpha^{h-\gamma}$ BCN($\alpha, \beta, \gamma, \gamma$)s when $h \geq \gamma$. Thus proved.   □

Theorem 3.8 *For any $n = \alpha + \beta$, the optimal α that maximizes the total number of servers in BCN($\alpha, \beta, \gamma, \gamma$) is given by*

$$
\alpha \approx (2 \cdot \gamma \cdot n)/(2 \cdot \gamma + 1).
\tag{3.8}
$$

Proof The total number of servers in BCN($\alpha, \beta, \gamma, \gamma$) is denoted as

$$
\begin{aligned}
f(\alpha) &= \alpha^\gamma \cdot (\alpha + \beta) \cdot (\alpha^\gamma \cdot \beta + 1) \\
&= n \cdot \alpha^\gamma + n^2 \cdot \alpha^{2\gamma} - n \cdot \alpha^{2\gamma+1}.
\end{aligned}
$$

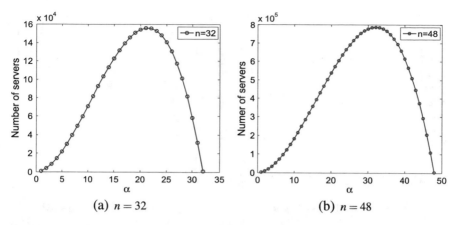

(a) $n = 32$ (b) $n = 48$

Fig. 3.5 The network order of BCN($\alpha, \beta, 1, 1$) versus α ranging from 0 to n

Thus, we have

$$\frac{\varphi f(\alpha)}{\varphi \alpha} = n \cdot \alpha^{\gamma-1}\left(\gamma + 2\gamma \cdot n \cdot \alpha^{\gamma} - (2\gamma + 1)\alpha^{\gamma+1}\right)$$

$$\approx n \cdot \alpha^{\gamma-1}\left(2\gamma \cdot n \cdot \alpha^{\gamma} - (2\gamma + 1)\alpha^{\gamma+1}\right).$$

Clearly the derivative is 0 when $\alpha \approx (2 \cdot \gamma \cdot n)/(2 \cdot \gamma + 1)$. At the same time, the second derivative is less than 0. Thus, $\alpha \approx (2 \cdot \gamma \cdot n)/(2 \cdot \gamma + 1)$ maximizes the total number of servers in BCN($\alpha, \beta, \gamma, \gamma$). Thus proved.   □

Figure 3.5 plots the number of servers in BCN($\alpha, \beta, 1, 1$) when $n = 32$ or 48. The network order goes up and then goes down after it reaches the peak point as α increases in the both cases. The largest network order of BCN($\alpha, \beta, 1, 1$) is 787,968 for $n = 48$ and 155,904 for $n = 32$, and can be achieved only if $\alpha = 32$ and 21, respectively. Such experimental results match well with Theorem 3.8.

Figure 3.6a depicts the changing trend of the ratio of the network order of BCN($\alpha, \beta, 1, 1$) to that of FiConn($n, 2$) as the number of ports in each mini-switch increases, where α is set to the optimal value $\alpha \approx (2 \cdot \gamma \cdot n)/(2 \cdot \gamma + 1)$. The results show that the number of servers of BCN is significantly larger than that of FiConn($n, 2$) with the server degree 2 and the network diameter 7, irrespective the value of n.

Formula 3.7 indicates that the network order of BCN grows double-exponentially when h increases from $\gamma - 1$ to γ, while grows exponentially with h in other cases. On the contrary, the network order of FiConn always grows double-exponentially with its level. Consequently, it is not easy to incrementally deploy FiConn because a level-k FiConn requires a large number of level-$(k - 1)$ FiConns. In the case of BCN, incremental deployment is relative easy since a higher level BCN requires only α one lower level BCNs except $h = \gamma$. On the other hand, the incomplete BCN can

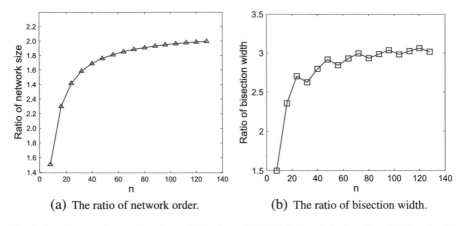

(a) The ratio of network order.   (b) The ratio of bisection width.

Fig. 3.6 The ratio of network order and bisection width of BCN(α, β, 1, 1) to that of FiConn(n, 2), where their network diameters are the same 7

relieve the restriction on the network order for realizing incremental deployment by exploiting the topological properties of BCN in both dimensions.

3.5.2   Low Network Diameter and Server Degree

According to Theorems 3.4 and 3.5, we obtain that the diameters of BCN(α, β, h) and BCN(α, β, h, γ) ($h < \gamma$) are $2^{h+1} - 1$ and $2^{\gamma+1} + 2^{h+1} - 1$, respectively. In practice, h and γ are two small integers. Therefore, BCN is a low-diameter network.

After measuring the network order and diameter of BCN, we study the node degree distribution in BCN(α, β, h, γ). If $h < \gamma$, the node degrees of master servers are 2 except the α available master servers for further expansion. The α master servers and all of the slave servers are of degree 1. Otherwise, there are $\alpha \cdot (\alpha^{\gamma} \cdot \beta + 1)$ available master servers that are of degree 1. Other master servers and all of the slave servers are of degree 2.

BCN of level one in each dimension offers more than 1,000,000 servers if 56-port switches are used, while the server degree and network diameter are only 2 and 7, respectively. This demonstrates the low network diameter and server degree of BCN.

3.5.3   Connectivity and Path Diversity

The edge connectivity of a single server is one or two in BCN(α, β, h, γ). Consider the fact that BCN(α, β, h, γ) is constituted by a given number of low level subnets in the first dimension. We further evaluate the connectivity of BCN at the level of different subnets in Theorem 3.9.

Theorem 3.9 *In any $BCN(\alpha, \beta, h, \gamma)$, the smallest number of remote links or servers that can be deleted to disconnect one $BCN(\alpha, \beta, i)$ from the entire network is*

$$\begin{cases} \alpha - 1, & \text{if } h < \gamma \\ \alpha - 1 + \alpha^i \cdot \beta, & \text{if } h \geq \gamma \end{cases} \tag{3.9}$$

Proof If $h < \gamma$, consider any subnet $BCN(\alpha, \beta, i)$ for $0 \leq i < h$ in $BCN(\alpha, \beta, h, \gamma)$. If it contains one available master server for further expansion, only $\alpha - 1$ remote links are used to interconnect with other homogeneous subnets. It is clear that the current subnet is disconnected if the corresponding $\alpha - 1$ remote links or servers are removed.

If $h \geq \gamma$, besides the $\alpha - 1$ remote links that connect its master servers the subnet $BCN(\alpha, \beta, i)$ has $\alpha^i \cdot \beta$ additional remote links that connect its slave servers. Thus, it can be disconnected only if the corresponding $\alpha - 1 + \alpha^i \cdot \beta$ remote links or servers are removed. Thus proved. $\qquad\square$

Theorem 3.10 *Bisection width: The minimum number of remote links that need to be removed to split $BCN(\alpha, \beta, h, \gamma)$ into two parts of about the same size is given by*

$$\begin{cases} \alpha^2/4, & \text{if } h < \gamma \text{ and } \alpha \text{ is an even integer} \\ (\alpha^2 - 1)/4, & \text{if } h \geq \gamma \text{ and } \alpha \text{ is an odd integer} \\ \alpha^{h-\gamma} \cdot \frac{(\alpha^\gamma \cdot \beta + 2) \cdot \alpha^\gamma \cdot \beta}{4}, & h \geq \gamma \end{cases} \tag{3.10}$$

Proof It is worth noticing that the bisection width of a compound graph $G(G_1)$ is the maximal one between the bisection widths of G and G_1 [14]. For $1 \leq h < \gamma$, $BCN(\alpha, \beta, h, \gamma)$ is a compound graph, where G is a complete graph with α nodes and G_1 is $BCN(\alpha, \beta, h - 1, \gamma)$. We can see that the bisection width of G is $\alpha^2/4$ if α is an even number and $(\alpha^2 - 1)/4$ if α is an odd number. The bisection width of $BCN(\alpha, \beta, h - 1, \gamma)$ can be induced in this way and is the same as that of G.

$BCN(\alpha, \beta, h, \gamma)$ for $h \geq \gamma$ is a compound graph, where G is a complete graph with $\alpha^\gamma \cdot \beta + 1$ nodes and G_1 is $BCN(\alpha, \beta, h)$. We can see that the bisection width of G is $(\alpha^\gamma \cdot \beta + 2) \cdot \alpha^\gamma \cdot \beta/4$ and that of G_1 is $\alpha^2/4$ if α is an even number and $(\alpha^2 - 1)/4$ if α is an odd number, which is less than that of G. Moreover, G_1 has $\alpha^{h-\gamma}$ copies of $BCN(\alpha, \beta, \gamma)$, and there is one link between the ith $BCN(\alpha, \beta, \gamma,)$s in two copies of G_1 for $1 \leq i \leq \alpha^{h-\gamma}$. Thus, there are $\alpha^{h-\gamma}$ links between any two copies of G_1; hence, the bisection width of $BCN(\alpha, \beta, h, \gamma)$ is $\alpha^{h-\gamma} \cdot (\alpha^\gamma \cdot \beta + 2) \cdot \alpha^\gamma \cdot \beta/4$. Thus proved. $\qquad\square$

For any $FiConn(n, k)$, the bisection width is at least $N_k/(4 * 2^k)$, where $N_k = 2^{k+2} * (n/4)^{2^k}$ denotes the number of servers in the network. We then evaluate the bisection width of $FiConn(n, 2)$ and $BCN(\alpha, \beta, 1, 1)$ under the same server degree, switch degree, and network diameter. Figure 3.6b shows that $BCN(\alpha, \beta, 1, 1)$ significantly outperforms $FiConn(n, 2)$ in terms of the bisection width. Larger bisection width implies higher network capacity and more resilient against failures.

As proved in Lemma 3.2, there are $\alpha - 1$ node-disjoint paths between any two servers in $BCN(\alpha, \beta, \gamma, \gamma)$, where the optimal α is $2 \cdot \gamma \cdot n/(2 \cdot \gamma + 1)$. Thus, the path diversity between any two servers is about $\lceil 2n/3 \rceil - 1$. With such disjoint paths, the transmission rate can be accelerated and the transmission reliability can be enhanced. Table 3.2 summarizes the network orders and path diversities of $BCN(\alpha, \beta, 1, 1)$ with different n. We see from Table 3.2 that $BCN(\alpha, \beta, 1, 1)$ has high path diversity for unicast traffic. We need to weigh between maximizing network size and maximizing path diversity. In fact, the path diversity reaches the highest level when $\alpha = n$.

3.5.4 Evaluation of the Path Length

We run simulations on $BCN(\alpha, \beta, 1, 1)$ and $FiConn(n, 2)$ in which $n \in \{8, 10, 12, 14, 16\}$ and α is set to its optimal value. The ratio of network order of BCN to that of FiConn varies between 1.4545 and 1.849. For the all-to-all traffic, Fig. 3.7a shows the average length of the shortest path of FiConn, the shortest path of BCN, and the routing path of BCN. For any BCN, the routing path length is a little bit larger than the shortest path length since the current routing protocols do not

Table 3.2 Network orders and path diversities of $BCN(\alpha, \beta, 1, 1)$

n	8	16	24	32	40	48
Network order	640	98,56	49,536	155,904	380,160	787968
Path diversity	5	10	15	21	26	31

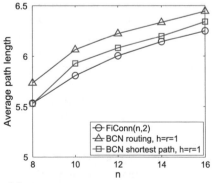
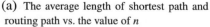
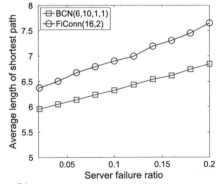

(a) The average length of shortest path and routing path vs. the value of n

(b) The average length of routing path vs. the server failure ratio.

Fig. 3.7 The path length of BCN and that of FiConn under different configurations

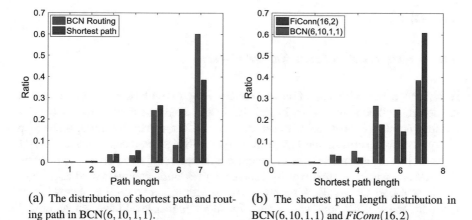

(a) The distribution of shortest path and routing path in BCN(6, 10, 1, 1).

(b) The shortest path length distribution in BCN(6, 10, 1, 1) and *FiConn*(16, 2)

Fig. 3.8 The path length distribution under all-to-all traffic

entirely realize the shortest path routing. The FdimRouting can be further improved by exploiting those potential shortest paths due to links in the second dimension. Although the network order of BCN is a lot larger than that of FiConn, the average shortest path length is a little bit larger than that of FiConn.

Then we evaluate the fault-tolerant ability of the topology and the routing algorithm of BCN(6, 10, 1, 1) and a FiConn(16, 2). The network sizes of BCN(6, 10, 1, 1) and FiConn(16, 2) are 5856 and 5327, respectively. As shown in Fig. 3.7b, the average routing path length of BCN and FiConn increase with the server failure ratio. The average routing path length of BCN is a lot shorter than that of FiConn under the same server failure ratio although FiConn outperforms BCN in terms of the average shortest path length when the server failure ratio is zero. Such results demonstrate that the topology and routing algorithms of BCN possess better fault-tolerant ability. Note that the BCN(6, 10, 1, 1) in Fig. 3.7b holds less servers than BCN(11, 5, 1, 1) in Fig. 3.7a.

To further compare the distribution characteristics of pat length in BCN(α, β, 1, 1) and FiConn(n, 2), we run simulations with the settings of $n = 16$ and $\alpha = 6$. The network sizes of BCN(6, 10, 1, 1) and FiConn(16, 2) are 5856 and 5327, respectively. The simulation results shown in Fig. 3.8 match the theoretical values of the network diameters of BCN and FiConn. Figure 3.8a indicates that the routing algorithm of BCN may not discover a few part of shortest paths, leading to longer routing paths to some extent. Figure 3.8b plots the distribution of shortest path for BCN and FiConn. The two network topologies have the same server degree 2, same network diameter 7, and the similar network order. However, their distributions of shortest path length are not the same. We can see that about 60% of the shortest paths in FiConn are of length 7 while only about 40% of the shortest paths in BCN are of length 7.

3.6   Discussion

3.6.1   Extension to More Server Ports

Although we assume that all of the servers are equipped with two built-in NIC ports, the design methodologies of HCN and BCN can be easily extended to involve any constant number, denoted as m, of server ports. In fact, servers with four embedded NIC ports have been available. Given any server with m ports, it can contribute $m - 1$ ports for future higher-level interconnection after reserving one port for connecting with a mini-switch. In other words, a server with m ports can be treated as a set of $m - 1$ dual-port servers. Each server connects with the same mini-switch using its first port, and uses the other $m - 1$ ports for future higher-level interconnections.

3.6.2   Locality-Aware Task Placement

HCN and BCN have many advantages, such as good topological properties, easy implementation and low cost. However, they cannot achieve the same aggregate capacity as Fat-tree and BCube under all-to-all traffic. The root cause is that the number of cables and switches in HCN and BCN are far less than that in other networks. Fortunately, such an issue can be addressed by some techniques at the application layer, due to the observations as follows.

As prior work [5] has shown, a server is likely to communicate with a small subset of other servers when conducting typical applications in common data centers, such as the group communication, and VM migration. Additionally, data centers with hierarchical network topologies, for example HCN and FiConn, hold an inherent benefit. That is, lower level networks support local communications, while higher level networks are designed to realize remote communications.

Furthermore, a locality-aware approach can be used for placing those tasks onto servers in HCN. That is, those tasks with intensive data exchange can be placed onto servers, in HCN$(n, 0)$, which connect to the same switch. If those tasks need some more servers, they may reserve a one higher lever topology HCN$(n, 1)$, and so on. There is only a few even one server hop between those servers. As proved above, HCN is usually sufficient to contain hundreds even thousands of servers, where the number of server hops is at most three. Therefore, the locality-aware mechanism can largely save the network bandwidth by avoiding unnecessary remote data communications.

3.6.3   Impact of the Server Routing

In both HCN and BCN, since servers that connect to other modules at a different level have to forward packets, they will need to devote some processing resources for this aspect. Although we can use software based packet forwarding schemes for our HCN and BCN, they usually incur non-trivial CPU overhead. The hardware-based packet forwarding engine like CAFE [15] and ServerSwitch, are good candidates for supporting DCN designs. Inspired by the fact that CAFE and ServerSwitch can be easily configured, we can re-configure them to forward self-defined packets for our HCN or BCN without any hardware re-designing.

References

1. Al-Fares M, Loukissas A, Vahdat A. A scalable, commodity data center network architecture [J]. ACM SIGCOMM Computer Communication Review, 2008, 38(4): 63–74.
2. Greenberg A, Hamilton J R, Jain N, et al. VL2: a scalable and flexible data center network [J]. ACM SIGCOMM Computer Communication Review, 2009, 39(4): 51–62.
3. Guo C, Wu H, Tan K, et al. Dcell: a scalable and fault-tolerant network structure for data centers [J]. ACM SIGCOMM Computer Communication Review, 2008, 38(4): 75–86.
4. Li D, Guo C, Wu H, et al. FiConn: Using backup port for server interconnection in data centers [C]. In Proc. of 28th IEEE INFOCOM, Rio de Janeiro, Brazil, 2009: 2276–2285.
5. Li D, Guo C, Wu H, et al. Scalable and cost-effective interconnection of data-center servers using dual server ports [J]. IEEE/ACM Transactions on Networking, 2011, 19(1): 102–114.
6. Guo C, Lu G, Li D, et al. BCube: a high performance, server-centric network architecture for modular data centers [J]. ACM SIGCOMM Computer Communication Review, 2009, 39(4): 63–74.
7. Wu H, Lu G, Li D, et al. MDCube: a high performance network structure for modular data center interconnection [C]. In Proc. of 5th ACM CoNEXT, Rome, Italy, 2009: 25–36.
8. Alon N, Hoory S, Linial N. The Moore bound for irregular graphs [J]. Graphs and Combinatorics, 2002, 18(1): 53–57.
9. Damerell R M. On Moore graphs [C]. In Proc. of Cambridge Philosophical Society, Cambridge, UK, 1973:227–236.
10. Imase M and Itoh M. A design for directed graphs with minimum diameter [J]. IEEE Transactions on Computers, 1983, 32(8): 782–784.
11. Agrawal D P, Chen C, Burke J R. Hybrid graph-based networks for multiprocessing [J]. Telecommunication system, 1998, 10:107–134.
12. Breznay P T, Lopez M A. Tightly connected hierarchical interconnection networks for parallel processors [C]. In Proc. of 22th IEEE ICPP, NY, USA, 1993, 1: 307–310.
13. Breznay P T, Lopez M A. A class of static and dynamic hierarchical interconnection networks [C]. In Proc. of 23th IEEE ICCP, Raleigh, USA, 1994, 1:59–62.
14. Agrawal D P, Chen C, Burke J R. Hybrid graph-based networks for multiprocessing [J]. Telecommunication Systems, 1998, 10(1-2): 107–134.
15. Lu G, Shi Y, Guo C, et al. CAFE: a configurable packet forwarding engine for data center networks [C]. In Proc. of 2nd ACM SIGCOMM workshop on Programmable routers for extensible services of tomorrow, 2009: 25–30.

Chapter 4
DCube: A Family of Network Topologies for Containerized Data Centers

Abstract There are two distinct trends in building large-scale data centers. The first trend is to interconnect massive servers directly through a scalable network topology, and the second trend is to build a large-scale data center by interconnecting a large number of data center modules. Within each data center module, thousands of servers are interconnected through a certain network topology, and are packed into a container, which serves as a basic building block. This chapter introduces a family of intra-module network topologies designed for modular data centers called DCube, including H-DCube and M-DCube. Each DCube interconnects a large number of dual-port servers and low-end switches. A large number of DCube modules can be further interconnected to form a new modular data center. The evaluation results show that the DCube exhibits a graceful performance degradation as the server or switch failure rate increases. Moreover, the DCube significantly reduces the required wires and switches compared to the BCube and fat-tree. The methodologies proposed in this chapter can apply to the compound graph of the basic building block and other hypercube-like graphs, such as Twisted cube, Flip MCube, and fastcube.

4.1 Introduction

One of the basic design goals of Data Center Networking (DCN) is to interconnect massive servers, according to a specific interconnection topology. There are two distinct trends in building large-scale data centers. The first trend is to interconnect massive servers directly through a scalable network topology. As described in Chap. 2, many types of network topologies have been proposed for large-scale data centers.

The second trend is to design large-scale distributed data centers, also known as modular data centers, on the basis of a large number of single data centers modules [1, 2]. In a data center module, thousands of servers are interconnected through a certain network topology, and are packed into a container. On the basis of many data center modules, the inter-module network topology is introduced to construct a larger data center. uFix [3] and MDCube [4] are two representative inter-module interconnection topologies. MDCube adopts BCube as its intra-module intercon-

© Springer Nature Singapore Pte Ltd. 2022
D. Guo, *Data Center Networking*,
https://doi.org/10.1007/978-981-16-9368-7_4

nection topology, and it uses optical fiber to connect high-speed switch ports of different modules directly. MDCube shows a solution to inter-module interconnection, while uFix focuses on how to interconnect heterogeneous data center modules. This modular construction method can greatly reduce the cost of data center construction, management and maintenance, while greatly improving the flexibility and convenience of the construction and deployment of data center.

This chapter introduces a family of low-cost and robust network topologies, called DCube(n, k), for modularized data centers with dual-port servers and low-end n-port switches. DCube(n, k) consists of k interconnected sub-networks, each of which is a compound graph made by interconnecting a certain number of basic building blocks by means of a hypercube-like graph (or its variant 1-möbius). In each subnetwork, the basic building block is just n/k servers connected to a switch. In this chapter, two kinds of DCube(n, k), H-DCube and M-DCube, are described. They are constructed based on compound graph theory, and adopt hypercube topology and 1-möbius topology, respectively. M-DCube is designed to further improve the aggregate bandwidth of H-DCube.

An n-dimensional hypercube possesses 2^n nodes and $n \times 2^{n-1}$ edges, and its network diameter is n. The address of each node is represented as $X = x_n \ldots x_i \ldots x_1$, in which the x_i of any dimension is binary variable. Two nodes are called mutual neighbors if and only if their addresses differ by one dimension. möbius topology is a variant of hypercube topology, and it inherits many properties of hypercube. For example, möbius and hypercube have the same number of nodes and edges, similar connectivity, regularity and iterative property. However, the network diameter of möbius is $\lceil (m + 1)/2 \rceil$, which is about half the diameter of the hypercube topology, and the average path length of möbius is about two-thirds of the hypercube.

Both of the above two DCube topologies show good regularity and symmetry, but they face challenges in efficiently interconnecting a large number of dual-port servers, because it is necessary to ensure that the interconnection topology has a smaller network diameter and a higher bisection bandwidth. In addition, DCube provides higher network bandwidth for the unicast transmission between any pair of servers, and has good fault-tolerant transmission capability. Mathematical analysis and simulation experiment results show that DCube significantly reduces the number of required wires and switches compared to BCube; hence, the construction cost, energy-consumption, and cabling complexity are largely reduced. In addition, DCube achieves a higher speedup than BCube for the unicast and multicast traffic patterns. A challenge that arises here is the fact that DCube cannot achieve the same aggregate bottleneck throughput (ABT) as BCube, which employs more ports for each server and switch for routing.

Note that the proposed methodologies in this chapter can be applied to the compound graph of the basic building block and other hypercube-like graphs, such as Twisted cube, Flip MCube, and Fastcube, and the resulted topologies are similar to DCube.

4.2 The DCube Topology

This section first discusses the core concepts of DCube topology, and then describes two typical topologies in DCube family, including H-DCube and M-DCube. These two topologies adopt hypercube and möbius respectively in the construction process based on the compound graph theory.

4.2.1 Design Idea of DCube

DCube network is built with two kinds of devices: dual-port servers and n-port mini-switches. The basic building block, denoted by Cube, is simply n servers connecting to an n-port mini-switch. After arranging the n servers into k groups, the Cube is partitioned into k sub-blocks, denoted by $Cube_0, Cube_1, \ldots, Cube_{k-1}$. Each sub-block is built with $m = n/k$ servers connecting to the n-port switch in the basic building block, as shown in Fig. 4.1. A DCube network consists of k sub-networks, denoted by $DCube_1, DCube_2, \ldots, DCube_k$, which share all of the mini-switches in the DCube network. Throughout this chapter, we impose a limitation on the value of k such that $n \bmod k = 0$.

For $0 \leq i \leq k - 1$, $DCube_i$ is a compound graph of $Cube_i$ and a hypercube-like graph. $DCube_i$ is obtained by replacing each node of the hypercube-like graph with a copy of $Cube_i$ and replacing each link of the hypercube-like graph with a link, which connects two corresponding copies of $Cube_i$. In $DCube_i$, the topology properties of the hypercube-like graph are preserved at the cost of an additional link that is associated with each server in $DCube_i$. The compound graph theory has been discussed in detail in Chaps. 2 and 3 of this book, and will not be repeated here. For each server in $DCube_i$, the first port is used to connect to the switch while the second port is used to interconnect with another server in a different copy of $Cube_i$. Although the construction of DCube requires that $DCube_i$s adopt the homogeneous

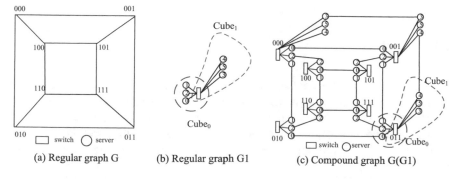

(a) Regular graph G   (b) Regular graph G1   (c) Compound graph G(G1)

Fig. 4.1 H-DCube with $n = 6$ and $k = 2$

hypercube-like graph, the basic ideas also apply to the heterogeneous setting. For example, the application of möbius cube is also studied in this chapter besides the standard hypercube topology.

When constructing a DCube$_i$, a *constraint* that arises is that the node degree of the hypercube-like graph must be equal to the number of servers in Cube$_i$, so as to constitute a regular compound graph. Thus, this requires an m-dimensional *hypercube-like* graph, in which each node is assigned a unique address $a_{m-1} \ldots a_1 a_0$. For $0 \leq i \leq k - 1$, we can infer that DCube$_i$ has $2^m \times m$ servers and 2^m switches; hence, DCube has $2^m \times m \times k = 2^m \times n$ servers and 2^m switches. We can see that DCube can be uniquely defined by two parameters, n and k, and is characterized by DCube(n, k). For ease of presentation, we use the term DCube to represent DCube(n, k) throughout the rest of this chapter.

We now present the construction of DCube(n, k) as follows. We number the k subnetworks from DCube$_0$ to DCube$_{k-1}$ and number all switches from 0 to $2^{n/k} - 1$. Equivalently, we use an address $a_{m-1} \ldots a_1 a_0$ to denote a switch. We can use a term u to number those servers that are connected to the same switch from 0 to $n - 1$, and we can denote a server in DCube(n, k) using the form $\langle a_{m-1} \ldots a_1 a_0, u \rangle$. The connection rule between servers using their second ports depends on the used m-dimensional hypercube-like graph. In this chapter, we focus on the hypercube of diameter m and the 1-möbius cube of diameter $\lceil (m + 1)/2 \rceil$. The resulting topologies are characterized by H-DCube and M-DCube, respectively. The basic ideas also apply to other hypercube-like graphs with a similar diameter as that of the 1-möbius cube, such as the 0-möbius cube, Twisted cube, Flip MCube, and Fastcube.

Before presenting the construction approach for the H-DCube and M-DCube, we first introduce notations and definitions used throughout this chapter.

1. Let e_j denote the m-dimensional binary vector with only the jth dimension equals to 1, where j is the index of e_j.
2. Let E_j denote the m-dimensional binary vector with 1 in dimensions x_j through x_0, where j is the index of e_j.
3. Given two m-dimensional binary vectors, + denotes the modulo-2 addition for the corresponding elements.

4.2.2 H-DCube

In an m-dimensional hypercube, denoted by $H(m)$, two nodes, $x_{m-1} \ldots x_1 x_0$ and $y_{m-1} \ldots y_1 y_0$, are called the mutual jth neighbors if their addresses differ by only the jth vector component. That is $y_{m-1} \ldots y_1 y_0 = x_{m-1} \ldots x_1 x_0 + e_j$, where $0 \leq j \leq m - 1$. The node degree and network diameter of $H(m)$ are well known to be m. In an H-DCube(n, k), any server $\langle a_{m-1} \ldots a_j \ldots a_0, u \rangle$ is interconnected with another server $\langle a_{m-1} \ldots \overline{a_j} \ldots a_0, u \rangle$ using their second ports, where $j = u \bmod m$. This simple connection rule guarantees the desired topology of an H-DCube network consisting of k sub-networks H-DCube$_i$ for $0 \leq i \leq k - 1$. We now discuss the correctness of such connection rule as in the following.

Only m servers and the unique switch in a basic building block falls into a sub-block Cube$_i$ if the sequence number u of those servers falls into the range of $[i \times m, (i + 1) \times m)$, where $0 \leq i \leq k - 1$. A sub-network, H-DCube$_i$, is a compound graph made by interconnecting a given number of copies of Cube$_i$ by means of $H(m)$, and is obtained by the following operations. Firstly, any node $\langle a_{m-1} \ldots a_j \ldots a_0 \rangle$ and its jth neighbor node $\langle a_{m-1} \ldots \overline{a_j} \ldots a_0 \rangle$ in $H(m)$ are replaced by two copies of Cube$_i$. Secondly, the link from node $\langle a_{m-1} \ldots a_j \ldots a_0 \rangle$ to its jth neighbor node in $H(m)$ is replaced by a remote link between servers $\langle a_{m-1} \ldots a_j \ldots a_0, u \rangle$ and $\langle a_{m-1} \ldots \overline{a_j} \ldots a_0, u \rangle$, where $u = i \times m + j$. It is easy to see that only one link is connected to the second port of each server and that method is equivalent to the aforementioned connection rule. We can infer that all k sub-networks can be constructed in the same way, and they share all of the 2^m switches. Thus, the connection rule can guarantee the desired topology of an H-DCube network.

Figure 4.1 plots an H-DCube with $n = 6$ and $k = 2$, which consists of 8 basic building blocks, each with 6 servers and one switch. Note that the entire topology of each of the three basic building blocks 000, 001, and 011 are plotted, while the topologies of the other basic building blocks are only partially plotted. All devices form two sub-networks, H-DCube$_0$ and H-DCube$_1$, each is a compound graph of Cube$_i$ and a 3-dimensional hypercube. The servers, whose sequence numbers are less than 3, belong to H-DCube$_0$, while others belong to H-DCube$_1$. Clearly, H-DCube$_0$ and H-DCube$_1$ share all of the switches in the network.

4.2.3 M-DCube

An m-dimensional möbius cube is such an undirected graph: its node set is the same as that of an m-dimensional hypercube; any node $X = x_{m-1} \ldots x_1 x_0$ connects to m other nodes Y_j $(0 \leq j \leq m - 1)$, where Y_j satisfies one of the following equations:

$$Y_j = \begin{cases} x_{m-1} \ldots x_{j+1}\overline{x_j}x_{j-1} \ldots x_0, & \text{if } x_{j+1} = 0 \\ x_{m-1} \ldots x_{j+1}\overline{x_j}\overline{x_{j-1} \ldots x_0}, & \text{if } x_{j+1} = 1 \end{cases} \qquad (4.1)$$

According to the above definition, a node X connects to its jth neighbor $Y_j = X + e_j$ that differs in bit x_j if $x_{j+1} = 0$ and to $Y_j = X + E_j$ if $x_{j+1} = 1$. The connection between X and Y_{m-1} has x_m as undefined. Here, x_m is either equal to 1 or 0, resulting in slightly different network topologies. This chapter assumes that $x_m = 1$: the resulting network is called the möbius cube. The node degree and network diameter of the m-dimensional 1-möbius cube are m and $\lceil (m + 1)/2 \rceil$, respectively.

In an M-DCube(n, k), all $2^m \times n$ servers and 2^m switches are first grouped into 2^m basic building blocks, each of which consists of n servers connecting to one switch using their first ports. For any server $\langle a_{m-1} \ldots a_{j+1}a_j a_{j-1} \ldots a_0, u \rangle$, we connect it to a

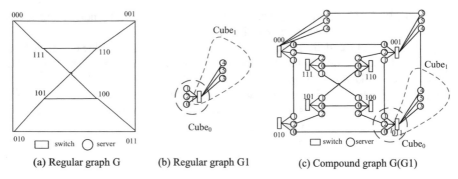

(a) Regular graph G   (b) Regular graph G1   (c) Compound graph G(G1)

Fig. 4.2   M-DCube with $n = 6$ and $k = 2$

server $\langle a_{m-1} \ldots a_{j+1} \overline{a_j} a_{j-1} \ldots a_0, u \rangle$ if $a_{j+1} = 0$ or $\langle a_{m-1} \ldots a_{j+1} \overline{a_j a_{j-1} \ldots a_0}, u \rangle$ if $a_{j+1} = 1$ via their second ports, where $j = u \bmod m$. This connection rule guarantees the desired topology of an M-DCube network consisting of k sub-networks M-DCube$_i$ for $0 \leq i \leq k - 1$. Figure 4.2 shows an M-DCube network with $n = 6$ and $k = 2$, which consists of 8 basic building blocks, each with 6 servers and one switch. The servers, whose sequence numbers are less than 3, belong to M-DCube$_0$ and others are associated with M-DCube$_1$.

In summary, we can support 2048 servers in DCube(8,1) using 8-port switches and 4096 in DCube(16,2) using 16-port switches for both H-DCube and M-DCube. Another possible way of using 16-port switches is to construct DCube(16,1) with 1048576 servers, which is too large for modular data centers. Moreover, the network diameter and expected routing path length are also relatively higher than that of DCube(16,2). Inspired by such fact, we prefer to construct DCube(n, k) as $k > 1$ interconnected sub-networks when the number of ports on each switch exceeds an upper bound, for example 8.

4.3   Single-Path Routing for the Unicast Traffic

The unicast traffic is the basic traffic pattern, which is the basic component of the incast and broadcast traffic. In this section, we focus on the single-path routing scheme for the unicast traffic pattern, which makes local decisions to identify a path or the next hop for any pair of servers based only on their addresses.

For two servers, A and B, we use $h(A, B)$ to denote the hamming distance between the two switches that are connecting the two servers, respectively, which is the number of different digits in their address arrays. It is clear that the maximum hamming distance between two switches in a DCube(n, k) is $m = n/k$. In this , two servers are neighbors if they connect to the same switch or if they directly connect to each other. The distance between two neighboring servers is one. Additionally, two switches are neighbors if there exists at least one pair of directly connected servers, each

---

Algorithm 1 H-DCubeRouting(A, B)

Require: A=$\langle a_{m-1} \ldots a_0, u_a \rangle$ and B=$\langle b_{m-1} \ldots b_0, u_b \rangle$
1: $path(A, B) = \{A, \}$;
2: $symbols$ is a permutation of **Expansion-hypercube**(A, B);
3: $Pswitch = \langle a_{m-1} \ldots a_0 \rangle$ and $Cswitch = \langle a_{m-1} \ldots a_0 \rangle$;
4: **while** $symbols$ not empty **do**
5:　　Let e_i denote the leftmost term in $symbols$;
6:　　$Cswitch = Cswitch + e_i$ and $u = \lfloor u_a/m \rfloor \times m + i$;
7:　　append $\langle Pswitch, u \rangle$ and $\langle Cswitch, u \rangle$ to $path(A, B)$;
8:　　remove e_i from $symbols$ and $Pswitch = Cswitch$;
9: append server B to $path(A, B)$;
10: return $path(A, B)$;

Expansion-hypercube(A, B)

1: $terms = \{\}$;
2: **for** $i = m - 1$ to 0 **do**
3:　　**if** $A[i] \neq B[i]$ **then**
4:　　　　　　　　　　　　　　　　　　　　　　$\triangleright A[i] = a_i$; $B[i] = b_i$.
5:　　　　append e_i to $terms$;
6:　　return $terms$;

---

belonging to one of the two switches. Actually, the construction rules of H-DCube and M-DCube ensure that two neighboring switches have k pairs of such connecting servers, and each belongs to one sub-network. For example, two switches, 000 and 001, are neighbors since two servers, $\langle 000, 0 \rangle$ and $\langle 000, 3 \rangle$, directly connect to two servers, $\langle 001, 0 \rangle$ and $\langle 001, 3 \rangle$, respectively, as shown in Fig. 4.1.

Based on such facts, we design two routing algorithms, H-DCubeRouting and M-DCubeRouting, as shown in Algorithms 4.1 and 4.3 respectively, to find a single path for any server pair.

4.3.1 Single-Path Routing in H-DCube

In H-DCubeRouting, we assume that A=$\langle a_{m-1} \ldots a_0, u_a \rangle$ and B=$\langle b_{m-1} \ldots b_0, u_b \rangle$ are the source and destination servers, respectively. We first find a sequence of switches by correcting one digit of the previous switch, so as to produce a switch path from the source switch $a_{m-1} \ldots a_0$ to the destination switch $b_{m-1} \ldots b_0$. To make two adjacent switches in the switch path be neighbors, we have to choose one from the k pairs of connecting servers, each is connected to one of the adjacent switches.

A natural way of selecting the pair of connected servers belonging to the same sub-network, H-DCube$_i$ ($i = \lfloor u_a/m \rfloor$), is shown in Algorithm 4.1. Generally, another pair of connecting servers is also desirable if the two servers belong to the same sub-network, H-DCube$_i$ ($i = \lfloor u_b/m \rfloor$), as the destination server B. These efforts ensure that all intermediate servers in a routing path belong to the same sub-network, so as to ensure the load balance of each server under a uniform traffic model. The switches in the resulting path of Algorithm 4.1 can be uniquely determined by the addresses of servers and hence are omitted from the path. From Algorithm 4.1, we obtain the Theorem 4.1.

Theorem 4.1 *The diameter of an H-DCube(n, k) is $2 \times m + 1$, where $m = n/k$.*

Proof In an H-DCube(n, k), the shortest path between any two servers traverses, at most, $m + 1$ switches, including the source switch, the destination switch, and other $m - 1$ intermediate switches. For any intermediate switch, there exists a one-hop packet transmission from the server receiving a packet to another server, which will forward the packet to its neighboring server in the next switch along with the switch path. For the source switch, there also exists a one-hop packet transmission if the source server cannot directly forward a packet to a server in the next switch. For the destination switch, a one-hop packet transmission is also necessary if the server receiving a packet is not the destination server. In addition, the total length of these m inter-switch sub-paths between any adjacent switches in the shortest path is m. Thus, Theorem 4.1 is proven.   □

4.3.2   Single-Path Routing in M-DCube

We first discuss the expansion techniques of a vector, which are fundamental to our detailed discussion on the routing of M-DCube. The set $R = \{e_j, E_j | 0 \le j \le m - 1\}$ forms a redundant basis for Z_2^m. Any vector X in Z_2^m can be expanded by R in the form:

$$X = \sum_{j=1}^{m-1} (\alpha_j e_j + \beta_j E_j), \tag{4.2}$$

with each $\alpha_j \in \{0, 1\}$ and $\beta_j \in \{0, 1\}$.

Definition 4.1 For a vector X, the set of e_j and E_j with non-zero coefficients in Eq. 4.2, denoted as $E(X)$, is called an expansion of the vector X. Any $t \in E(X)$ is a term of this expansion of X. The weight of an expansion $E(X)$ is called $W(X)$ and is equal to the cardinality of $E(X)$.

There can be more than one expansion of a vector due to the use of a redundant basis. Thus, an expansion with minimal weight is referred to as the minimal expansion of X. Algorithm 4.2 shows a simple procedure for finding the minimal expansion for any vector. In each round, the algorithm first generates a sub-vector starting from the bit position $index$ to the rightmost bit position of the vector X. If the sub-vector is 1, a term E_0 is added into the $symbols$ set. If the sub-vector is 0, the algorithm is terminated. If the leftmost bit of the sub-vector is 0, the algorithm decreases the $index$ by one and executes the next round. If the leftmost two bits of the sub-vector are 10, a term e_{index} is appended to the $symbols$ set. Otherwise, a term E_{index} is added into the $symbols$ set, and the vector X is updated by $X + E_{index}$ since the

---

Algorithm 2 Expansion-mobius(A)

Require: $A = a_{m-1} \ldots a_0$ is a m-dimensional vector over $\{0, 1\}$; $A[i] = a_i$
1: $symbols = \{\}$ and $index = m - 1$;
2: **while** $index < 0$ **do**
3:     **if** $index == 0$ **then**
4:         **if** $A[index] == 1$ **then**
5:             append E_0 to $symbols$;
6:             $index = index - 1$;
7:     **else**
8:         **if** $A[index]==0$ **then**
9:             $index==index - 1$;
10:         **else**
11:             **if** $A[index]A[index - 1] == 10$ **then**
12:                 append e_{index} to $symbols$;
13:             **if** $A[index]A[index - 1] == 11$ **then**
14:                 append E_{index} to $symbols$;
15:                 $A = a_{m-1} \ldots \overline{a_{index-2}} \ldots a_0$;
16:             $index = index - 2$;
17: return $symbols$;

---

term E_{index} complements all bits from the position $index$ to the rightmost position of X. The algorithm then carries out the next round after decreasing the $index$ by two.

For a source server $A = \langle a_{m-1}a_{m-2} \ldots a_0, u_a \rangle$ and a destination server $B = \langle b_{m-1}b_{m-2} \ldots b_0, u_b \rangle$ in M-DCube(n, k), we define $A + B$ as the vector obtained by the mod 2 sum of the switch addresses $a_{m-1}a_{m-2} \ldots a_0$ and $b_{m-1}b_{m-2} \ldots b_0$. To generate the shortest path between A and B, we first derive a switch path from the source switch $a_{m-1}a_{m-2} \ldots a_0$ to the destination switch $b_{m-1}b_{m-2} \ldots b_0$. We then find a pair of servers to connect two adjacent switches indirectly. Actually, the switch path between any pair of switches in M-DCube(n, k) is equivalent to the path between two corresponding nodes in the m-dimensional möbius cube.

For any switch, e_i or E_i denotes its immediate neighbor along dimension i. For this reason, we refer to e_i or E_i as a routing symbol. To form a switch path, a sequence of routing symbols should be applied to the source switch. The minimal expansion $E(A + B)$, achieved by Algorithm 4.2, cannot be directly used to produce the switch path due to the following challenging issue. According to the definition of a 1-möbius cube, given any node, only one of e_i and E_i can be the routing symbol along the ith dimension, where $0 \leq i \leq m - 1$. Consequently, a routing symbol in the minimal expansion does not always correspond to an edge in the 1-möbius cube, and hence may be inapplicable to the current node. A natural way to deal with this issue is to replace any inapplicable routing symbol with an equivalent routing sequence obtained from Theorem 4.2.

Theorem 4.2 *Given a node $A = a_{m-1}a_{m-2} \ldots a_0$:*

1. if e_i is inapplicable to the node A, it can be replaced by an equivalent routing sequence, $E_i E_{i-1}$ or $E_{i-1} E_i$, which is applicable to A.

2. *if E_i is inapplicable to the node A, it can be replaced by an equivalent routing sequence, $e_i E_{i-1}$ or $E_{i-1} e_i$, which is applicable to A.*

Proof It is clear that $e_i = E_i + E_{i-1}$ and that $E_i = e_i + E_{i-1}$. Assume that e_i is inapplicable to node A. This implies that $a_{i+1} = 1$; hence, E_i is applicable to node A. If $a_i = 1$, then E_{i-1} is applicable to node A; thus, $E_{i-1} E_i$ is applicable to node A. Here, $E_i E_{i-1}$ is inapplicable to node A since traversal along edge E_i from node A makes a_i become 0. If $a_i = 0$, E_{i-1} is inapplicable to node A, but the application of E_i complements the bit a_i. Now E_{i-1} is applicable to node $A + E_i$, making $E_i E_{i-1}$ be applicable to node A.

Assume that E_i is inapplicable to node A. This implies that $a_{i+1} = 0$; thus, e_i is applicable to node A. If $a_i = 1$, then E_{i-1} is applicable to node A; thus, $E_{i-1} e_i$ is applicable to node A since traversal along edge E_{i-1} from node A does not complement the bit a_{i+1}. If $a_i = 0$, the application of e_i complement bit a_i. Now E_{i-1} is applicable to node $A + e_i$, making $e_i E_{i-1}$ be applicable to A. Thus, Theorem 4.2 is proven.   □

According to the aforementioned strategies, we design M-DCubeRouting, as shown in Algorithm 4.3, to find a path from a source server A to a destination server B. The algorithm begins with achieving the minimal expansion of $A + B$ by invoking Algorithm 4.2. It then calls the exact-routing algorithm to derive a sequence of routing symbols, which can be successfully applied to the source switch so as to establish a switch path to the destination switch. In each round, the exact-routing algorithm ranks all terms in the *symbols* set in descending order according to the

---

Algorithm 3 M-DCubeRouting(A, B)

Require: A=$\langle a_{m-1} \ldots a_0, u_a \rangle$ and B=$\langle b_{m-1} \ldots b_0, u_b \rangle$;
1: *symbols*=**Expansion-mobius**($A + B$);
2: $path(A, B) = \{A\}$;
3: **Exactrouting**($a_{m-1} \ldots a_0$, *symbols*);
4: append server B to $path(A, B)$;

Exactrouting (S, *symbols*)

Require: S denotes a current switch in the shortest path;
1: **while** *symbols* is not empty **do**
2:     Let t denote the leftmost term in *symbols*;
3:     **if** t is applicable to S. **then**
4:         Let t' denote the rightmost applicable term to S in *symbols*; ▷ The rightmost applicable term can be the leftmost applicable term.
5:         $u = \lfloor u_a / m \rfloor \times m + i$, where i is the index of $t' = e_i$ or $t' = E_i$;
6:         append $\langle S, u \rangle$ and $\langle S + t', u \rangle$ to $path(A, B)$;
7:         remove t' from *symbols* and **Exact-routing**($S + t'$, *symbols*);
8:     **else**
9:         **if** The term t is in the form e_i **then**
10:             replace e_i with $E_i E_{i-1}$ if $a_i = 0$ or $E_{i-1} E_i$ otherwise;
11:         **else**
12:             replace E_i with $e_i E_{i-1}$ if $a_i = 0$ or $E_{i-1} e_i$ otherwise;
13:         **Exactrouting**(S, *symbols*);

---

index of each term and then examines the leftmost term. If the leftmost term t is inapplicable to the current switch S, it is replaced by the equivalent routing sequence defined in Theorem 4.2. If the leftmost term is applicable to the current switch S, the rightmost applicable term t' will be applied first and then it updates the current switch S and *symbols*. This strategy can avoid the appearance of the worst result as in the following. If the leftmost applicable term is E_j, the application of it will make makes all next applicable terms become inapplicable.

After deriving the shortest switch path between the source and destination servers, we need to choose one from k pairs of the connecting servers between any adjacent switch S and $S + t'$ so as to make them be neighbors in M-DCube(n, k). As shown in Fig. 4.2, two switches, 000 and 111, are neighbors since servers, $\langle 000, 2 \rangle$ and $\langle 000, 5 \rangle$, are directly connected to servers, $\langle 111, 2 \rangle$ and $\langle 111, 5 \rangle$, respectively. It is natural to choose the pair of servers which belong to the same sub-network, M-DCube$_i$ $(i = \lfloor u_a/m \rfloor)$, as the source server A. Servers $\langle S, u \rangle$ and $\langle S + t', u \rangle$ append to the routing path, where $u = \lfloor u_a/m \rfloor \times m + i$ and i denotes the index of t'. Actually, another pair of connecting servers is also desirable if they belong to the same sub-network, M-DCube$_i$ $(i = \lfloor u_b/m \rfloor)$, as the destination server B. From Algorithm 4.3, we obtain the following theorem.

Theorem 4.3 *The diameter of an M-DCube(n, k) is $2 \times \lceil (m + 1)/2 \rceil + 1$, where* $m = n/k$.

Proof Given any two servers, A and B, in an m-dimensional 1-möbius cube, the weight of minimal expansion $E(A + B)$ is at most $\lceil m/2 \rceil$ since no two terms have adjacent indices, according to Algorithm 4.2. The leftmost inapplicable term t_i in the minimal expansion is then replaced by a routing sequence with a length of 2. This strategy ensures that no other inapplicable terms exist in the routing path after replacing t_i since E_{I-1} complements the bit a_{j+1} for any inapplicable term t_j, where $j < i$. Thus, Algorithm 4.3 ensures that the diameter of an m-dimensional 1-möbius cube is $\lceil (m + 1)/2 \rceil$, and hence the shortest path between any two servers in an M-DCube(n, k) traverses, at most, $\lceil (m + 1)/2 \rceil + 1$ switches. As mentioned in the proof of Theorem 4.1, there is a one-hop transmission within each switch and between two adjacent switches in the routing path. Thus, Theorem 4.3 is proven. □

4.4 Multi-paths Routing for the Unicast and Multicast Traffic

Traditionally, two parallel paths between a source server and a destination server exist if the intermediate servers and switches on one path do not appear on the other. It is clear that there exists, at most, two parallel paths for any pair of servers under this strict definition due to the dual-port on each server. In this chapter, we relax this definition slightly. That is, two paths are called *parallel* if the intermediate switches on one path are not involved in the other path, except for the beginning and ending switches. In addition, two neighboring switches possess k pairs of directly connected servers. To maximize the utility of such an advantage, a switch path can be utilized as k *weak parallel paths*, which share the same set of switches but have different intermediate servers. Such parallel and weak parallel paths between any pair of servers can be further utilized to improve the transmission rate or to enhance the transmission reliability for the unicast traffic with Multipath TCP. In addition, Multipath TCP can explore such multiple paths to tackle traffic congestion, leading to higher network utilization.

The following theorem specifies the exact number of parallel paths and weak parallel paths between any two servers in a DCube(n, k).

4.4.1 Multi-path Routing for the Unicast Traffic in H-DCube

Theorem 4.4 *There are m parallel and n weak parallel paths between any two servers in an H-DCube(n, k), where $m = n/k$.*

The m parallel paths between any two servers in an H-DCube(n, k) can be simplified to m parallel switch paths since all inter-switch sub-paths of two adjacent switches in the m paths are disjoint. Thus, we can show the correctness of Theorem 4.4 by constructing such m parallel switch paths. Algorithm 4.1 produces a shortest switch path from A to B using any permutation of the minimal expansion $E(A + B)$, which contains e_j for some $0 \le j \le m - 1$ but not E_j for any $0 \le j \le m - 1$. In the minimal expansion of $A + B$, $W(A + B)$ distinct terms form an initial routing sequence, resulting in $W(A + B)!$ minimal routing sequences. Theorem 4.5 indicates that only $W(A + B)$ parallel switch paths from A to B can be generated.

Theorem 4.5 *Let the minimal expansion $E(A + B)$ generated by the* Expansion-hypercube *be $t_1, t_2, \ldots, t_{W(A+B)}$. Algorithm 4.1 generates $W(A + B)$ parallel switch paths from A to B using permutations as follows. The ith permutation for $0 \le i < W(A + B)$ is denoted as $p_1, p_2, \ldots, p_{W(A+B)}$, where $p_j = t_{(j+i) \bmod W(A+B)}$ for $1 \le j \le W(A + B)$.*

Proof Actually, these permutations are obtained by moving each term of the initial routing sequence to the mod left by i, for $0 \le i < W(A + B)$, under the following two constraints. Firstly, any pair of such permutations differ in the addition of leftmost j terms for $1 \le j \le W(A + B)$. Secondly, the addition of any leftmost j terms is different from that of any left j' terms where $j \ne j'$ for each of these permutations. Thus, this pattern ensures that the resulting $W(A + B)$ paths are disjoint except for the source and destination switches; thus, the $W(A + B)$ parallel switch paths are produced. For example, $e_1 e_0$ and $e_0 e_1$ are two minimal routing sequences, resulting in two parallel switch paths between servers $\langle 000, 0 \rangle$ and $\langle 011, 0 \rangle$, as shown in Fig. 4.1. The resulting two parallel paths are $\{\langle 000, 0 \rangle, \langle 000, 1 \rangle, \langle 010, 1 \rangle, \langle 010, 0 \rangle, \langle 011, 0 \rangle\}$ and $\{\langle 000, 0 \rangle, \langle 001, 0 \rangle, \langle 001, 1 \rangle, \langle 011, 1 \rangle, \langle 011, 0 \rangle\}$, respectively.

Assume that t'_h belongs to $\{e_{m-1}, \ldots, e_1, e_0\}$ but does not appear in the minimal expansion $E(A + B)$. We achieve a new routing sequence by appending t'_h to the leftmost and rightmost terms of one existing routing sequence. This further results in a switch path, which is parallel with the $W(A + B)$ switch paths generated in Theorem 4.5. For example, $e_2 e_1 e_0 e_2$ or $e_2 e_0 e_1 e_2$ produces another path, which is parallel with the two paths generated by $e_1 e_0$ and $e_0 e_1$, for two servers, $\langle 000, 0 \rangle$ and $\langle 011, 0 \rangle$, as shown in Fig. 4.1. The path generated by $e_2 e_1 e_0 e_2$ is $\{\langle 000, 0 \rangle, \langle 000, 2 \rangle, \langle 100, 2 \rangle, \langle 100, 1 \rangle, \langle 110, 1 \rangle, \langle 110, 0 \rangle, \langle 111, 0 \rangle, \langle 111, 2 \rangle, \langle 011, 2 \rangle, \langle 011, 0 \rangle\}$. The path resulting from $e_2 e_0 e_1 e_2$ is $\{\langle 000, 0 \rangle, \langle 000, 2 \rangle, \langle 100, 2 \rangle, \langle 100, 0 \rangle, \langle 101, 0 \rangle, \langle 110, 1 \rangle, \langle 111, 1 \rangle, \langle 111, 2 \rangle, \langle 011, 2 \rangle, \langle 011, 0 \rangle\}$.

Consider that $m - W(A + B)$ terms in $\{e_{m-1}, \ldots, e_1, e_0\}$ do not appear in the minimal expansion $E(A + B)$. Thus, we can derive $m - W(A + B)$ parallel switch paths with lengths of $W(A + B) + 2$ using the same approach as mentioned above. Thus, we can construct m parallel switch paths between two servers, A and B in an H-DCube(n, k). If we produce another switch path between A and B using a new routing sequence, at least one switch in the new path has to have appeared on existing switch paths. The root cause for this is that the leftmost and rightmost terms of the new routing sequence must have to have appeared at the beginning and/or end of the m previous routing sequences. Thus, the largest number of parallel switch paths between any pair of servers in an H-DCube(n, k) must be m.

After discussing the parallel switch paths between any pair of servers in an H-DCube(n, k), we further consider the sub-path between any adjacent switches in these paths. Algorithm 4.1 selects one pair of connecting servers for each pair of neighboring switches in any switch path so as to realize a path including servers and switches. However, k weak parallel paths can be produced based on a given switch path between two servers after updating line 6 with $u = j \times m + i$ for $0 \leq j \leq k - 1$. That is, each one of the m parallel paths between two servers can be realized as k weak parallel paths. For this reason, we can induce that there are $m \times k = n$ weak parallel paths between any two servers. Thus, Theorem 4.4 is proven. □

4.4.2 Multi-path Routing for the Unicast Traffic in M-DCube

Theorem 4.6 *There are m parallel and n weak parallel paths between any two servers in an M-DCube(n, k), where m = n/k.*

Proof We use a similar approach to show the correctness of Theorem 4.6 by constructing such parallel paths and weak parallel paths. Given two servers, A and B, in an M-DCube(n, k), *Expansion-mobius* generates a minimal expansion of $A + B$ in the scenario of an m-dimensional möbius cube. It is worth noticing that some terms in the minimal expansion may be inapplicable to the current switch and should thus be replaced by an equivalent routing sequence as defined in Theorem 4.2. To address this issue, *M-DCubeRouting* generates an initial routing sequence by invoking *exact-routing* with the minimal expansion $E(A + B)$ as input. Assume that the initial routing sequence is denoted as $t_1, t_2, \ldots, t_l$, where $l \geq W(A + B)$.

M-DCubeRouting can further generate some parallel switch paths between servers A and B by using permutations of the initial routing sequence. Any permutation on the initial routing sequence forms a new routing sequence. For this reason, one can conclude that there exists $l!$ routing sequences, but only the following ones can produce l parallel switch paths. Assume that the ith permutation for $0 \leq i < l$ is denoted as $p_1, p_2, \ldots, p_l$, where $p_j = t_{(j+i) \bmod l}$ for $1 \leq j \leq l$.

The first challenging issue we face is the fact that terms in each permutation of the initial routing sequence may be inapplicable to the current switch and should be revised according to Theorem 4.2 so as to generate an applicable routing sequence. The resulting applicable routing sequence can generate a new switch path, which will be parallel with existing switch paths. For example, the initial routing sequence for a shortest path from server $A = \langle a_2 a_1 a_0 = 000, 0 \rangle$ to server $B = \langle 100, 0 \rangle$ in Fig. 4.2 is $E_2 E_1$, which is applicable. The first permutation of $E_2 E_1$ is it. The second permutation of $E_2 E_1$ is $E_1 E_2$, in which E_1 is inapplicable to $A = 000$ since $a_2 = 0$. As a result, $E_1 E_2$ should be replaced by $e_1 E_0 E_2$.

Besides the l parallel switch paths, we will show how generate other $m - l$ parallel switch paths between any servers, A and B, in an M-DCube(n, k). Let t'_m denote any term, which belongs to $\{e_{m-1}, \ldots, e_1, e_0\}$, but is not the leftmost term in the routing sequences defined by the above permutation operation. We achieve a new routing sequence by appending t'_m to the leftmost and rightmost terms of one existing routing sequence, which further results in a new switch path. This switch path is parallel to the l switch paths generated by the aforementioned l permutations of the initial routing sequence. For example, the routing sequence, $e_0 E_2 E_1 e_0$, is achieved by appending $t'_m = e_0$ to the beginning and end of $E_2 E_1$. It then generates a new path from $A = \langle 000, 0 \rangle$ to $B = \langle 100, 0 \rangle$ in Fig. 4.2. That is, $\{\langle 000, 0 \rangle, \langle 001, 0 \rangle, \langle 001, 2 \rangle, \langle 110, 2 \rangle, \langle 110, 1 \rangle, \langle 101, 1 \rangle, \langle 101, 0 \rangle, \langle 100, 0 \rangle\}$.

The second challenging issue we face is the fact that appending a t'_m term to the leftmost and rightmost terms of an existing routing sequence, for example the initial routing sequence $t_1, t_2, \ldots, t_l$, does not necessarily result in a parallel path in a general scenario. Actually, if t'_m appears at the end of an existing routing sequence, then the last two switches, including the destination switch, in the new switch path must have occurred in the related path.

To produce a parallel path based on $t'_m t_1, t_2, \ldots, t_l t'_m$, we need to find a t''_m from $t_1, t_2, \ldots, t_l$, which does not appear at the end of those routing sequences defined by the above permutation operation. If there exists such a t''_m, we move it to the end of $t'_m t_1, t_2, \ldots, t_l t'_m$ and optimize it so as to generate an applicable and parallel path. Otherwise, we need to replace a given term in $t_1, t_2, \ldots, t_l$ with an equivalent routing sequence, as defined in Theorem 4.2, which contains such a t''_m. We then move the t''_m to the end of the resulting routing sequence and optimize it so as to generate an applicable and parallel path. Based on those techniques, we can derive other $m - l$ switch paths which are parallel with the l switch paths generated by the above permutation operation.

As discussed in the proof of Theorem 4.4, there are k pairs of connected servers between any two neighboring switches. For this reason, M-DCubeRouting produces k weak parallel paths, based on one switch path between two servers, by updating line 5 with $u = j \times m + i$ for $0 \le j \le k - 1$. Thus, one can induce that the m parallel paths between two servers can be realized as $m \times k = n$ weak parallel paths; hence, Theorem 4.6 is proven.   □

4.4.3 Speedup for the Multicast Traffic

A complete graph consisting of a set of servers can speed up data replications in distributed file systems. We show that edge-disjoint complete graphs with $m + 1$ servers can be efficiently constructed in a DCube(n, k).

Theorem 4.7 *In a DCube(n, k), a server $\langle src, u_s \rangle$ and a set of m servers can form an edge-disjoint complete graph, where each of the m servers connects to a different neighboring switch of the switch src.*

In the case of H-DCube(n, k), the ith neighbor of switch src is defined as $src + e_i$ for $0 \le i < m$. Assume that $src + e_i$ and $src + e_j$ are two neighboring switches of the switch src, where $i \ne j$. A switch path with a length of two from $src + e_i$ to $src + e_j$ can be generated by a routing sequence, $e_j e_i$, and is denoted as $\{src + e_i, src + e_i + e_j, src + e_i + e_j + e_i = src + e_j\}$. It is easy to see that two different pairs of e_i and e_j cannot produce the same result of $e_i + e_j$, where $i \ne j$. Consequently, this pattern ensures that the switch paths among a switch src and its m neighbors are edge-disjoint.

In the case of M-DCube(n, k), the ith neighbor of switch src is defined as $src + t_i$ for $0 \le i < m$, where t_i is e_i or E_i according to the definition of an m-dimensional möbius cube. Assume that $src + t_i$ and $src + t_j$ are two neighbors of the switch src, where $i \ne j$. A switch path from $src + t_i$ to $src + t_j$ can be generated by an initial routing sequence, $t_j t_i$. In special cases, the resulting switch path is applicable and denoted as $\{src + t_i, src + t_i + t_j, src + t_i + t_j + t_i = src + t_j\}$. In general cases, each term in the initial routing sequence might be inapplicable. To generate an applicable switch path, each inapplicable term should be replaced with an equivalent routing sequence consisting of two terms, as defined in Theorem 4.2. For this reason, we can induce that the applicable and shortest switch path from $src + t_i$ to $src + t_j$ is, at most, four hops. In addition, we can see that two different pairs of t_i and t_j cannot produce the same result of $t_i + t_j$, where $i \ne j$. That is, the switch paths among a switch src and its m neighbors are edge-disjoint.

From the above construction approaches, we can see that the resulting complete graph is only two switch hops and is, at most, four switch-hops in H-DCube(n, k) and M-DCube(n, k), respectively.

Given an edge-disjoint complete graph formed by the source switch src and its m neighboring switches, each edge in the complete graph should be replaced by a pair of connected servers since two adjacent switches are not connected directly. It is worth noticing that there are k pairs of connecting servers for each edge in the complete graph. We only choose the pair of servers, which are located in the same sub-network DCube$_i$ as the source server, $\langle src, u_s \rangle$. The motivation is to separate the traffic in k complete graphs for any server $\langle src, u_s \rangle$ into the corresponding sub-networks. This operation ensures that the whole paths among the source server and m selected servers, including switches and servers, are still edge-disjoint.

We further show how to choose the m servers, denoted as d_j for $0 \le j \le m - 1$, for the source server $\langle src, u_s \rangle$. For the jth neighboring switch of the switch src, we choose d_j from n servers connecting to that switch, such that d_j locates in the same sub-network DCube$_i$ as the source server, where $i = \lfloor u_s/m \rfloor$. In this way, each d_j has m choices since a switch allocates m of n servers to each sub-network DCube$_i$. In this chapter, we just randomly select d_j from m choices and, we will study other selection methods of d_j in our future work.

So far, we have demonstrated that a complete graph can be formed by the source server and a set of m selected servers in the sub-network DCube$_i$. Actually, we can generate k such complete graphs for any server $\langle src, u_s \rangle$ by the following approach since a DCube(n, k) consists of k sub-networks DCube$_i$s. Assume that the selected server for d_j in the sub-network DCube$_i$, is denoted as $\langle src + t_j, u_j \rangle$ for $0 \le j \le m - 1$. The corresponding server $\langle src + t_j, u_j + k \times m \rangle$ and the source server generate a new complete graph in the kth sub-network DCube$_k$, where $k \ne i$.

A file on distributed file systems can be divided into chunks, and each chunk is typically replicated to three chunk servers. The source and the chunk servers establish a pipeline to reduce the replication time, as discussed in literature [5]. The edge-disjoint complete graph that is built into DCube works well for chunk replication speedup. When one writes a chunk to r ($r \le m + 1$) chunk servers, it sends $1/r$ of the chunk to each chunk server. Meanwhile, every chunk server distributes its copy to the other $r - 1$ servers by using the edge-disjoint edges. Consequently, this will be r times faster than the pipeline model.

4.5   Analysis and Evaluation

In this section, we conduct simulations to evaluate several basic properties of DCube. They include the speedup for the unicast and multicast traffic patterns, aggregate bottleneck throughput based on measurements of real-world data center traffic from literature [6], the cost, the power consumption, and the cabling complexity. We also compare the performance of DCube with not only Fat-tree but also DCell, HCN, Fat-tree and BCube, which are three particularly enlightening server-centric data-center structures. In the evaluation setting, the number of servers in DCube ranges from 2048 to 12288, and the capacity of each link is 1Gb/s. The setting matches the scale and configurations of a typical modular data center.

To ensure a fair comparison, such network topologies interconnect the same number of servers, denoted as N, with switches each of n ports. They, however, differ in the number of server ports, the number of switches, the number of cables, and the interconnection rules. DCell, HCN, and BCube are recursively defined structures, whose levels are denoted by k_1, k_2 and k_3, respectively, where $k_1 \leq k_3$.

4.5.1 Speedup for the Unicast and Multicast Traffic

For the unicast and multicast traffic patterns, we show the speedup as compared with other networking structure. We first summarize the throughput of such two traffic patterns under different networking topologies in Table 4.1.

For any server pair, A and B, DCube(n, k) provides $\lceil n/k \rceil$ parallel and n weak parallel paths for them. These properties not only speedup the unicast traffic, but also offer graceful degradation of performance. We can see from Fig. 4.3a that DCube offers more parallel paths for any pair of servers than HCN and BCube, as the network size increases from 2048 to 4096, 8192, and 12,288. Although DCell possesses more parallel paths for any server pair than DCube, DCube delivers large number of weak parallel paths and hence achieves better speedup performance for the unicast traffic.

For any source server, we show that the complete graph can significantly speedup the multicast traffic. Assume that server $A = \langle 000, 2 \rangle$ in Fig. 4.1 replicates 20G data

Table 4.1 Comparison of M-DCube, DCell, HCN, Fat-tree, and BCube

Throughput	M-DCube	DCell	HCN	Fat Tree	BCube
One to one	2	$k_1 + 1$	2	1	$k_3 + 1$
One to several	$\frac{n}{k} + 1$	$k_1 + 2$	n	1	$k_3 + 2$
All-to-all	$\frac{N}{1/3 \times n/k}$	$\frac{N}{2^{k_1}}$	$\frac{N}{2/3 \times 2^{k_2}}$	N	N

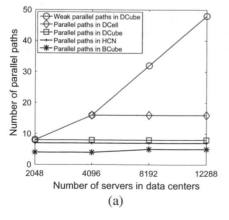
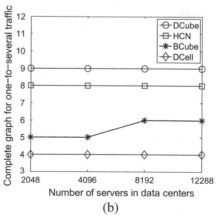

Fig. 4.3 **a** Number of parallel paths and weak parallel paths. **b** Order of the complete graph for the multicast traffic

to two servers, $B = \langle 010, 2 \rangle$ and $C = \langle 001, 2 \rangle$. With the complete graph approach, the data is split into two parts and sent to both B and C, respectively. B and C then exchange their data with each other. On the contrary, with the pipeline approach, A sends the data to B, and B sends the data to C. The complete graph can achieve about 2 times the speedup compared to the pipeline approach. In general, when a source deliver a chunk to r servers in the same complete graph, it sends $1/r$ of the chunk to each of the server. Meanwhile, every chunk server distributes its copy to the other $r - 1$ servers using the disjoint edges in the complete graph. This will be r times faster than the pipeline model. This implies that DCube executes speedup well when it comes to the multicast traffic pattern.

Recall that DCube can offer the largest complete graph of size $n/k + 1$. The largest cardinality of a complete graph in DCell and BCube is $k_1 + 2$ and $k_3 + 2$, as proved in literature [5]. Fig. 4.3b plots the largest cardinality, $r + 1$, of a complete graph for multicast traffic in DCube, DCell, HCN and BCube. We can see that DCube always outperforms others when the data center size increases from 2048 to 4096, 8192, and 12,288. This means that DCube results in a higher speedup than DCell, HCN and BCube for the multicast traffic pattern.

4.5.2   Aggregate Bottleneck Throughput

Aggregate bottleneck throughput (ABT) is defined as the number of flows times the throughput of the bottleneck flow under the all-to-all traffic pattern [7]. ABT of Fat-Tree and BCube are N since they achieve the nonblock communication between any pair of servers. ABT of Dcell is N/k_1, as proved in [7], while that of HCN is $\frac{N}{2/3 \times 2^{k_2}}$. ABT of H-DCube and M-DCube are proved in Theorem 4.8 and Lemma 4.1.

Theorem 4.8 *For a H-DCube(n, k) network, its ABT under the all-to-all traffic pattern is $\frac{N}{2/3 \times n/k}$, where n is the number of ports per switch and N is the number of servers.*

Proof The diameter of H-DCube(n, k) is $\frac{2 \times n}{k} + 1$, and the average path length approximates to n/k. The links in H-DCube(n, k) consist of two parts. Firstly, each of N server connects to a switch using its first port and thus generates one link. The number of such links is N. Secondly, each of N server connects with another server using its second port and thus generates one link. The number of such links is $N/2$. The total number of links in H-DCube(n, k) is $3N/2$. The number of flows carried in one link is

$$f_{num} = \frac{N(N-1)n/k}{3N/2},$$

where $N(N-1)$ is the total number of flows. The throughput one flow receives is thus $1/f_{num}$, assuming that the bandwidth of a link is one. The aggregate bottleneck throughput is therefore $N(N-1)\frac{1}{f_{num}} = \frac{N}{2/3 \times n/k}$. Thus, Theorem 4.8 is approved.   □

Fig. 4.4 ABT of different topologies under number of servers in data centers

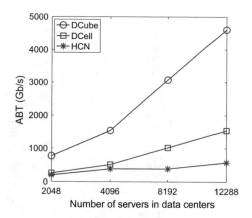

Lemma 4.1 *For a M-DCube(n, k) network, its ABT under the all-to-all traffic pattern is $\frac{N}{1/3 \times n/k}$, where n is the number of ports per switch and N is the number of servers.*

Proof The diameter of M-DCube(n, k) is $2 \times \lceil (n/k + 1)/2 \rceil + 1$, as we have proved in Theorem 4.3. Accordingly, we can derive that the expected distance approximates to $n/2k$. The proving process of Lemma 4.1 is similar to that of Theorem 4.8. □

It can be seen that the ABT of Fat-Tree and BCube is better than that of M-DCube in all-to-all traffic, since Fat-Tree and BCube use more switches, links and server ports than M-DCube. However, not all servers are frequently involved in all-to-all traffic, and the design of Fat-Tree and BCube will bring high switch costs or wiring costs. Besides, M-DCube can obtain higher ABT than HCN and DCell, because $n/(3k)$ is obviously less than $2^{k_2+1}/3$ and 2^{k_1}. In addition to the above theoretical analysis, we also conduct large-scale simulations to evaluate the ABT of the three data center interconnection structures. As shown in Fig. 4.4, M-DCube always gets higher ABT than HCN and Dcell with the change of data center size.

4.5.3 Qualification of Cost and Cabling Complexity

We first consider five networking topologies for a container with 2048 servers and many 8-port switches. Such topologies are constructed as follows. DCube topology is a DCube(8,1). DCell is a partial DCell(8, 2) with 28 DCell(8, 1)s. HCN topology is a partial HCN(8, 3) with 4 full HCN(8, 2)s. BCube topology is a partial BCube$_3$ with 4 full BCube$_2$s, where $n = 8$. Fat-tree topology has five layers of switches, with layers 0 to 3 having 512 switches per-layer and layer-4 having 256 switches. In this setting, DCube, DCell, HCN, BCube and Fat-tree employ 256, 252, 256, 1280 and 2304 8-port switches, while the number of NIC ports on each server are 2, 3, 4, 4, and 1, respectively. Note that a 8-port switch costs about $40 and consumes near 4.5W

		Cost(k$)		Power(kW)		Wires	Switchs
		Switch	NIC	Switch	NIC	No.	No.
2048	Fat-tree	92	10	10	10	10240	2304
	BCube	51	41	5.8	20	8192	1280 (8-port)
	HCN	10	20	1.2	15	3068	256 (8-port)
	DCell(2016)	10	40	1.1	18	4032	252 (8-port)
	DCube(8,1)	10	20	1.2	15	3072	256 (8-port)
4096	BCube	81	82	9.3	40	16384	2048 (8-port)
	HCN	20.5	82	2.3	41	6140	512 (8-port)
	BCube	115	61	16	37	12288	768 (16-port)
	DCell(4080)	38	81.6	5.4	41	8160	255 (16-port)
	DCube(16,2)	38	41	5.4	31	6144	256 (16-port)
8192	BCube	324	205	37.1	100	40960	8192 (8-port)
	HCN	41	164	4.6	82	12290	1024 (8-port)
	BCube	845	164	118	82	32768	5632 (16-port)
	DCell(8160)	76.5	163	10.7	82	16320	510 (16-port)
	DCube(32,4)	102	82	19.2	61	12288	256 (32-port)
12288	BCube	571	307.5	65	120	61440	14336 (8-port)
	HCN	61	246	7	123	18428	1536 (8-port)
	BCube	960	246	134	123	49152	6400 (16-port)
	DCell(12240)	114.7	245	16	122	24480	765 (16-port)
	DCube(48,6)	153.6	123	26	92	18432	256 (48-port)

Fig. 4.5   Measuring networking topologies under different sizes of data centers

of power. For one-port, two-port, and 4-port NICs, their costs are about $5, $10, and $20, while the power consumptions are about 5W, 7.5W, and 10W, respectively.

We then consider DCube, DCell, HCN and BCube for a container with 4096, 8192, and 12,288 servers, respectively. In this setting, DCube topologies are DCube(16,2) using 16-port switches, DCube(32,4) using 32-port switches, and DCube(48,6) using 48-port switches, respectively. DCell topologies are partial DCell(16,2)s with 15, 30 and 45 DCell(16,1)s using 16-port switches, respectively. HCN topologies using 8 ports switches are a HCN(8,3), a partial HCN(8,4) with 2 HCN(8,3)s, and a partial HCN(8,4) with 3 HCN(8,3)s, respectively. BCube topologies with 8-port switches are a full BCube$_3$, a partial BCube$_4$ with 2 full BCube$_3$s, and a partial BCube$_4$ with 3 full BCube$_3$s, respectively. Note that BCubes with 4096, 8192, and 12,288 servers may have different structures. For example, BCube topologies with 16-port switches are a full BCube$_2$, a partial BCube$_3$ with 2 full BCube$_2$s, and a partial BCube$_3$ with 3 full BCube$_2$s, respectively. Note that a 16-port switch costs about $150 and consumes 21W of power, a 32-port switch costs about $400 and consumes 75W of power, while a 48-port switch costs about $600 and consumes 103W of power.

Figure 4.5 summarizes the number of wires and switches, the cost of switches and NICs, and the power consumption of switches and NICs in such five networking topologies with 2048, 4096, 8192, and 12,288 servers, respectively. When DCell, HCN and BCube topologies are incomplete under some aforementioned settings, we

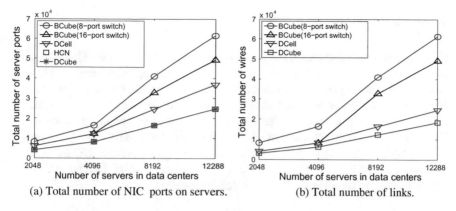

(a) Total number of NIC ports on servers.   (b) Total number of links.

Fig. 4.6   The sum of NIC ports and links vary as the increasing servers in data centers

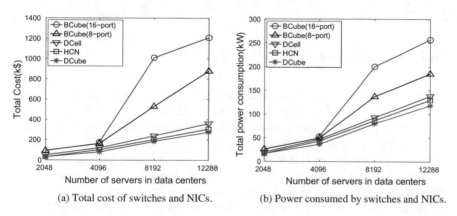

(a) Total cost of switches and NICs.   (b) Power consumed by switches and NICs.

Fig. 4.7   The cost and power consumption of switches and NICs under different settings

need to build partial structures. For a partial BCube$_i$, Guo et. al suggest that we build
the needed BCube$_{i-1}$s and then connect the BCube$_{i-1}$s using full layer$_i$ switches [5].

More precisely, Fig. 4.6 demonstrate that both the number of NIC ports on servers
and that of links of DCube are always considerably less than that of BCube and DCell,
irrespective the data center size. Thus, DCube largely reduces the cabling complexity,
especially for large containerized data centers. Note that DCube and HCN achieve
the similar performance in terms of the two metrics since they interconnect dual-
port servers. Moreover, DCube has other advantages due to the less number of wires
and switches. As shown in Figs. 4.5 and 4.7, DCube outperforms BCube, DCell
and HCN in terms of the entire cost and power consumption of switches and NICs,
irrespective the data center size. Additionally, we find that the BCube topology with
16-port switches results in more cost and power consumption compared to the BCube
topology with 8-port switches under the same number of servers.

4.5.4 Summary

DCube significantly reduces the required wires and switches compared to Fat-tree and BCube, because of this, it largely reduces the cabling complexity compared to other structures, especially for large containerized data centers. On the other hand, DCube considerably outperforms BCube in terms of the entire cost and power consumption, irrespective the data center size. Besides these benefits, the maximum throughput of DCube is twice that of Fat-tree, but less than that of DCell and BCube whose number of levels is typically larger than 2. For the multicast traffic pattern, DCube achieves a higher speedup than Fat-tree, DCell and BCube. Additionally, DCube achieve the higher ABT than DCell and offers graceful degradation.

4.6 Discussions

4.6.1 Locality-Aware Task Placement

Although the proposed network topologies have many advantages, such as easy wiring and low cost, they may not be able to achieve the same ABT as BCube under the all-to-all traffic pattern. Recall that BCube offers many NIC ports for each server and utilizes large numbers of switches so as to achieve a higher ABT; hence, significantly bringing about more cost and power consumption. In fact, the relatively lower ABT of DCube compared to BCube results from the lower number of links and switches, resulting in a longer average routing path; this is the tradeoff of other measurements. Fortunately, this issue can be addressed by some techniques at the application layer since a server is likely to communicate with a small subset of other servers for typical applications in common data centers.

Therefore, a locality-aware approach can be used for placing those tasks onto servers in DCube. That is, those tasks with intensive data exchanges can be first placed onto servers that connect to the same switch. If those tasks need some more servers, they may reserve the several nearest basic building blocks. There are only a few switch-hops, maybe even one, between those building blocks. It is easy to see that DCube is usually sufficient enough to contain hundreds of servers where the number of switch hops is, at most, two. Therefore, the locality-aware mechanism can largely save network bandwidth by avoiding unnecessary remote data communications.

4.6.2 Extension to More Server Interfaces

The basic idea of this chapter is to design a family of network topologies for containerized data centers using constant number of embedded NIC interfaces. Although we assume that all servers are equipped with two built-in NIC interfaces, the design methodologies of DCube can be easily extended to involve any constant number, denoted as q, of server interfaces. In fact, servers with four embedded NIC interfaces have been made available due to the rapid innovation on server hardware.

Consider that a set of $q - 1$ servers each of which holds two ports and connects with the same switch using its first port. It is clear that the set of $q - 1$ servers can totally contribute $q - 1$ ports for for interconnecting other basic building blocks. Given any server with q interfaces, it can contribute $q - 1$ interfaces for interconnecting other basic building blocks after reserving one port for connecting to switch. Intuitively, a server with q ports can be treated as a set of $q - 1$ dual-port servers. In this way, we can extend DCube to embrace any constant number of server ports. Additionally, each server can contribute its $q - 1$ interfaces as a virtual interface by the port trunking [4], e.g., three 1Gbps interfaces can be bundled into a virtual interface at 3Gbps. In this way, the link capacity between two servers can be significantly improved.

4.6.3 Impact of Server Routing

Server routing is a challenging issue faced by server-centric networking topologies for data centers. In a DCube structure, servers connecting to other basic building blocks have the responsibility of forwarding packets. Although DCube can use software-based or FPGA-based forwarding schemes, just as initial server-centric topologies do, it incurs forwarding delay. To further reduce the delay due to server routing, existing switches have been extended with dedicated devices like Server-Switch [8] and Sidecar [9, 10]. They integrate the programmable commodity switching chip into a built-in NIC for packet forwarding and leverage the CPU and RAM of server for in-network packet processing and storage, as prior work have shown [4, 8]. In such settings, a commodity server acts as not only an end host but also a mini-switch.

References

1. Hamilton J. Architecture for modular data centers [J]. arXiv preprint cs/0612110, 2007, 306–313.
2. Waldrop M M. Data center in a box [J]. Scientific American, 2007, 297(2): 90–93.
3. Li D, Xu M, Zhao H, et al. Building mega data center from heterogeneous containers [C]. In Proc. of 19th IEEE ICNP, Vancouver, BC Canada, 2011: 256–265.
4. Wu H, Lu G, Li D, et al. MDCube: a high performance network structure for modular data center interconnection [C]. In Proc. of 5th ACM CONEXT, Rome, Italy, 2009: 25–36.
5. Guo C, Lu G, Li D, et al. BCube: a high performance, server-centric network architecture for modular data centers [J]. ACM SIGCOMM Computer Communication Review, 2009, 39(4): 63–74.

6. Benson T, Akella A, Maltz D A. Network traffic characteristics of data centers in the wild [C]. In Proc. of ACM SIGCOMM Conference on Internet Measurement Conference, Melbourne, Australia, 2010.
7. Li D, Guo C, Wu H, Tan K, Zhang Y et al. Scalable and cost-effective interconnection of data-center servers using dual server ports [J]. IEEE/ACM Transactions on Networking, 2011, 19(1):102–114.
8. Lu G, Guo C, Li Y, et al. ServerSwitch: A Programmable and High Performance Platform for Data Center Networks [C]. In Proc. of USENIX NSDI, BOSTON, USA, 2011.
9. Costa P. Bridging the gap between applications and networks in data centers [J]. Operation System Review. 2013, 47 (1):3–8.
10. Alan S, Srikanth K, Gn S E. Sidecar: Building Programmable Datacenter Networks without Programmable Switches. In Proc. of HotNets, Monterey, CA, USA, 2010.

Chapter 5
R3: A Hybrid Network Topology for Data Centers

Abstract In large-scale data centers, many servers are interconnected via a dedicated network topology to satisfy specific design goals, such as the low equipment cost, the high network capacity, and the incremental expansion. The topological properties are critical factors that dominate the performance of the entire data center. The existing network topologies are either fully random or completely structured. Although such topologies exhibit advantages in given aspects, they suffer obvious shortcomings in other essential fields. In this chapter, we describe a general topology design methodology for data centers via the compound graph theory. We further propose a hybrid topology, called R3, which is the compound graph of a structured topology and a random topology. More precisely, R3 employs a random regular graph as a unit cluster and connects many such clusters by means of the generalized hypercube. R3 combines the advantages of the random regular graph and generalized hypercube, while effectively avoiding the shortcomings of the above two topologies.

5.1 Introduction

The basic goal of data center networking (DCN) is to effectively interconnect a large number of servers according to a specific network topology. As described in Chap. 2, many network topologies have been proposed for data centers from the industrial and academic communities, and these topologies can be divided into five categories based on their technological systems. In this chapter, the existing DCN structures are discussed from the aspect of their topology characteristics, and they are divided into two categories. The first is structured topologies, each of which impose strict interconnection rules on switches inside a data center. Fat-Tree [1], VL2 [2], BCN [3] fall into this category. The second is random topologies, each of which breaks the strict interconnection rules by introducing random links among switches, e.g., SMDC [4], Jellyfish [5], Scafida [6]. Structured topologies exhibit high throughput but are not incremental expansion. Those random topologies are incremental expansion, however, they suffer the complex cabling and routing processes. Actually, the superiority of structured and random topologies is complementary. In this chapter, the following

© Springer Nature Singapore Pte Ltd. 2022
D. Guo, *Data Center Networking*,
https://doi.org/10.1007/978-981-16-9368-7_5

two fundamental problems motivate us to design new networking topologies, which integrate their superiority and abandon their weakness.

Question 1: Can existing structured topologies of data centers satisfy the requirements of today's applications perfectly? For structured topologies, the strict interconnection rules simplify the construction of DCNs; however, they limit the incremental expansion of deployed data centers. In reality, a production data center needs to be gradually extended along with the increasing demands of applications and users. But the structured DCNs fail to support the incremental expansion.

Question 2: Can random topologies truly improve the performance of DCNs? Random DCNs support the incremental expansion naturally. Meanwhile, random links decrease the network diameter by connecting remote nodes. Furthermore, random DCNs can integrate heterogeneous devices during the expansion process. However, routing and maintenance in random DCNs are essential issues besides the cabling cost due to remote random links. First, the Dijkstra or Floyd algorithm, whose computation complexity is $O(n^2)$ or even $O(n^3)$, is used to search the shortest paths between any pair of nodes in random DCNs. It is clear that routing in random DCNs is time-consuming. Second, disordered and unsystematic link distribution incurs nontrivial maintain costs. That is, running a random DCN is relatively expensive.

Through the analysis of the above two types of network topologies, it is clear that the superiorities of structured DCNs and random DCNs are complementary. This motivates us to seek new topologies which can integrate their superiorities together and avoid their weakness. For this reason, we propose a family of hybrid topologies, which are compound graphs of given structured and random topologies. That is, given the number of random structures are unit clusters, which are then interconnected by means of a structured structure. In this way, a family of hybrid DCNs is built when utilizing a different random and structured graph. Consequently, the resulting hybrid DCNs can naturally integrate incremental expansion and fast routing characteristics via compound graph theory. Note that any of such hybrid topologies is a structured topology from the global viewpoint, while is a random topology from a local viewpoint.

5.2   The Design Methodology of Hybrid Topologies

In this chapter, we aim to combine the advantage of both structured topologies and random ones via designing a family of hybrid topologies for data centers. *Compound graphs* (either complete or incomplete) are introduced as the medium between random and structured graphs. More precisely, we combine the generalized hypercube (GHC) with the random regular graph (RRG) to derive a hybrid topology called *R3*. Since the concepts of generalized hypercube and compound graph are described in previous chapters, we just introduce the concept of the random regular graph in this chapter.

A random r-regular graph (r-RRG) is a graph selected from $G_{n,r}$, which denotes the probability space of all r-regular graphs on n vertices, where $3 \leq r < n$ and

$n \times r$ is even [7]. Generally, an RRG has excellent topology characteristics. First, all nodes have the same degree. Second, given r and n, an RRG has a lower bound on the network diameter. That is, for the same n, the network diameter of different RRGs is roughly $\log_{r-1} n$. r-RRGs have excellent properties such as coloring and the Hamiltonian cycle. Most importantly, RRGs support the incremental expansion properly by adding racks one by one without changing the existing network too much. Because of these outstanding characteristics, RRGs were introduced into data center topologies. Table 5.1 counts the important symbols and notations in this chapter.

5.2.1   Overview of Network Topologies

The existing wired DCNs topologies mainly belong to two categories, i.e., random ones and structured ones. Structured topologies lack scalability and incremental expansion, while the random ones suffer considerable routing overhead. For example, the routing in the hypercube is easy to implement since the hamming distance between two node identifiers will judge the existence of a direct link between any pair of nodes. When a hypercube-based data center needs to be extended, the data center has to double the number of servers. For Scafida, the topology can be incrementally expanded by adding any number of servers. Still, routing is very tough due to a large number of random links and the lack of topology information.

In this chapter, we pursue hybrid topologies for data centers, combining the superiorities of both random and structured topologies. The available structured topologies including Tree, Hypercube, Generalized Hypercube, Torus, and so on. The available random topologies including Small-world networks, Scale-free networks, Random regular graphs and so on. Given a structured topology and a random topology, we can use the construction method of the compound graph to build a hybrid network

Table 5.1 Symbols and notations

Term	Definition	Term	Definition
G	A simple graph	T	Total number of servers in R3
N	The total number of node in a graph	p	Port count of each switch in R3
$G(G_1)$	A compound graph based on G and G_1	t	Total switch that used in R3
m_i	Order of the ith dimension in GHC	α	# of switch ports that link with switches
Δ	Maximum node degree in a simple graph	β	# of switch ports that link with servers
$x'(G)$	Rdge chromatic number of G		

topology, as discussed in Chap. 3. Specifically, each node in the structured topology is replaced by a complete random topology called a random cluster. Each link in the structured topology is replaced by a link that connects two corresponding random clusters. Note that, in the resultant hybrid structure, each node represents a switch. Each switch uses parts of its ports to connect with other switches, and its remaining ports are used to connect all servers within the same rack.

In the resultant topology, the links derived from the structured graph are called **structured links**, while the links inside a random cluster are called **random links**. In our design, only one structured topology can be used since it is the container of random clusters, but heterogeneous random clusters can be used. The incremental expansion and routing will be demonstrated later.

5.2.2   R3: A Hybrid Topology Based on the Compound Graph Theory

As depicted above, the compound graph is a powerful method to integrate two kinds of topologies while remaining their superiorities. This motivates us to use the compound graph theory to construct our hybrid topologies. Different combinations of structured and random topologies result in different hybrid topologies. This will enlarge the design space and increase the design flexibility. In this way, our hybrid topologies enable designers to construct their data center networks on-demand.

However, to construct a hybrid topology correctly based on the compound graph theory, the following three constraints must be satisfied: (1) All random clusters must be interconnected via a structured topology; (2) The number of random clusters must equal to the number of chosen nodes in the structured topology; (3) The lower bound on the number of nodes in each random cluster can't be less than the maximum degree plus one in the structured topology.

Algorithm 1 Building Hybrid Topology, H

Require: Given a structured topology G, r denotes the number of nodes in G. Let $Adjacent[r][r]$ denote the adjacent matrix of G and R_i denote the random clusters.

1: Initialize each random cluster;
2: Let $Link[x]$ count all links already connected to the xth node in G;
3: Let $Degree[x]$ be the degree of the xth node in G;
4: **for** $i = 0$ to r **do**
5:    **for** $j = i$ to r **do**
6:       **if** $Adjacent[i][j] == 1$ and $Link[a] < Degree[a]$, $a = i, j$ **then**
7:          add a link between i^{th} and j^{th} random clusters;
8:          $Link[i]$++;
9:          $Link[j]$++;
10: **return** The hybrid topology H.

We describe the basic process of building a hybrid topology in Algorithm 5.1. In Algorithm 5.1, given the number of links that have been linked to each node in G, the adjacent matrix determines whether current nodes need to be linked. If ith and jth random clusters are connected, a link will be added to connect one random node in each cluster. If the number of nodes in a random cluster is less than the degree of nodes in the structured topology, then the degree of some nodes will be increased by more than one. Each proposed hybrid topology is the generalization of the involved random and structured topologies under specific settings. This design methodology has two extreme cases. If the random clusters have only one node, the resultant topology is just the structured graph we used, such as a 2D Torus depicted in Fig. 5.1a. If the structured graph is just a single node, then the hybrid topologies degrade into a fully random cluster, such as a random regular graph depicted in Fig. 5.1b.

Figure 5.1c depicts an example of hybrid topology constructed with the Torus and random topologies. If each random cluster is viewed as a node in Torus, then the topology is structured and hence easy-routing. While, in each random cluster, the topology is random and incremental expandable. In this hybrid topology, different random clusters are allowed to be embedded; meanwhile, it is not necessary that the number of nodes in such random clusters is the same. That is, this methodology can derive all hybrid topologies that lie between structured and random topologies.

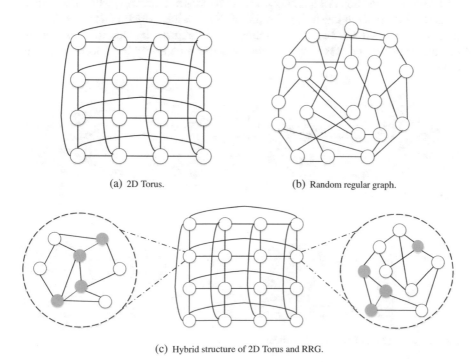

(a) 2D Torus.

(b) Random regular graph.

(c) Hybrid structure of 2D Torus and RRG.

Fig. 5.1 Structured and random topologies are special cases of the general topology

As aforementioned, the vast design space results in variable hybrid topologies with diverse randomness. To simplify the presentation, we focus on a representative hybrid topology, called *R3*. R3 employs the generalized hypercube GHC as its structured part and the random regular graph RRG as the unit of random clusters. Generalized hypercube and random regular graph are two representative topologies, which are widely used for designing network structures for data centers.

Definition 5.1 $R3(G(m_s, m_{s-1}, \ldots, m_1), r\text{-RRGs})$ denotes a kind of hybrid topologies, each of which is the compound graph of a random regular graph r-RRG and a GHC, whose dimensions are $m_s, m_{s-1}, \ldots, m_1$, respectively.

Definition 5.2 In a R3 topology, the selected nodes from each random cluster for linking with other random clusters are called *Boundary Nodes*.

Therefore, according to Definitions 5.1 and 5.2, a R3 topology contains number of $m_s \times m_{r-1} \times \cdots \times m_1$ RRGs and the amount of boundary nodes is $\sum(m_i - 1)$, where $1 < i < s$. Figure 5.2 depicts a 2×4 dimension topology, where eight random clusters of 3-RRGs are embedded, and each cluster accommodates 8 nodes. These structured links across random clusters form a 2×4 generalized hypercube. Every cluster is assigned a two-dimensional identifier (the two-digit red number in Fig. 5.2). Similarly, each node in a random cluster also owns its identifier. The allocation rules will be discussed later. All boundary nodes are chosen randomly, while the number of nodes in a random cluster can be an arbitrary value, which is not less than the node degree in GHCs. In R3, all parameters are adjustable. Due to the flexibility of parameter setting, R3s with different scales can be easily built according to its definition.

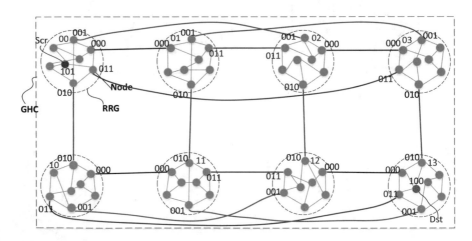

Fig. 5.2 R3(G(2,4),3-RRG) hybrid structure

5.2.3 Deployment Strategy for Data Centers with a Hybrid Topology

To put the hybrid topologies into real deployment, we investigate that both top-down wiring strategy and down-top wiring strategy are feasible.

The top-down deployment strategy mainly includes two steps. At the first step, we build the structured topology since it is the main skeleton of our hybrid topology. Having selected an appropriately structured topology, we deploy the boundary nodes of each cluster and interconnect these boundary nodes with structured links. When all structured links have been cabled, the first step will be terminated. At the second step, we fill the random clusters via adding nodes into the clusters in a way the random topologies require. In this way, the hybrid topology will be constructed successfully.

Unlike the top-down strategy, the down-top deployment strategy establishes the random clusters at the first step so that all random blocks are prepared for later interconnection. However, how to establish the random clusters depends on which topology the designer utilized. Note that the number of random clusters must be equal to the number of nodes in the structured topology we employed according to the compound graph theory. At the second step, we chose the boundary nodes from each random cluster and interconnected them together via structured links in a way the structured topology requires. When all links have been cabled, the construction process will be terminated.

5.3 Efficient Routing Methods of R3

Routing in R3 needs a dedicated design method because of the coexistence of random links and structured links. In hybrid topologies, the major challenge of routing comes from the embedded random clusters. In random topologies, like Scafida or Jellyfish, the shortest routing path between any pair of nodes can be decided only by Dijkstra-like methods, which incur considerable searching costs in large-scale data centers. In structured topologies, the topological characteristics can significantly ease the computation process of the shortest routing path.

To efficiently enable the routing, we propose an edge coloring-based routing algorithm to generate a path between any pair of nodes. An identifier is usually introduced to identify each node in existing DCNs. The identifier consists of two parts. Specifically, the inter-identifier contains the construction information of structured topology, while the inner-identifier locates nodes in each random cluster. The inter-identifier is determined by the behind structured topology. Regular links of the hybrid structure have different colors, and two nodes have the same inner-identifier if they are connected by a regular link. Therefore, inter-cluster routing paths can be inferred from inter-identifiers and colors of regular links, while the inner-cluster path can be generated through the Dijkstra algorithm. In addition, given the number of servers and switch ports, we can construct and optimize the model of the average shortest path. Moreover, we design the incremental expansion method at the cluster level, which supports the on-demand expansion of data centers.

Unlike BCube, where regularity and symmetry of the topology support fast routing under different flow patterns [8], the routing in R3, however, turns to be complicated due to the following two challenges. First, the unstructured topology of each random cluster denies the possibility of improving the efficiency of existing routing algorithms. Second, the associated nodes of each structured link can't be located precisely since they are chosen randomly. In this chapter, we focus on addressing the second challenging issue. It is clear that the obstacle of routing results from all of the random nodes. We thus regularize those random nodes by coloring all structured links and making routing just like in a structured topology.

Definition 5.3 An edge coloring of a graph is an assignment of colors to the edges of the graph so that no two adjacent edges have the same color. The least amount of employed colors is called the edge chromatic number, denoted as $x'(G)$ [9].

Theorem 5.1 *Let Δ denote the maximum node degree in a simple graph G, where $\Delta \leq x'(G) \leq \Delta + 1$. If $x'(G) = \Delta$, the graph is called class 1 graph, else, class 2 graph [9].*

Theorem 5.2 *Let K_n be the n-regular graph, then*

$$x'(G) = \begin{cases} n - 1, & if \ n \ is \ oven \\ n, & if \ n \ is \ odd \end{cases} \tag{5.1}$$

Theorems 5.1 and 5.2 have been proved in literature [9]. These theorems bring us an insight to identify boundary nodes in each random cluster. In addition, the structured topologies, e.g., Ring, Torus, Hypercube, Generalized Hypercube and Cayley Graph, are all of class 1 graphs.

The above theorems and description manifest that structured topologies that used in today's DCNs are colorable. Figure 5.2 depicted the color result on $R3(G(2, 4, 3\text{-}RRGs)$. To enable the success of edge coloring, the proposed Constraint 3 must be satisfied. After coloring each structured link, we assign each boundary node an inner-identifier according to the color of the associated link so that the nodes linked by the same link will have the same inner-identifier. In this novel way, the random cluster level paths will be calculated easily. So, the used inner-identifier of boundary nodes can significantly reduce the routing complexity.

5.3.1   Edge Coloring Based Identifier Allocation

To efficiently enable the routing, an identifier is usually introduced to identify each node in existing DCNs. In our hybrid topologies, the identifier consists of two parts.

The inter-identifier contains the construction information of structured topology, while the inner-identifier locates nodes in each random cluster.

The behind structured topology determines the inter-identifier. For example, if the structured topology is a Tesseract (4-dimension hypercube) that accommodates eight nodes, then a three-binary digit identifier can identify each node. If the structured topology is a $3 \times 4 \times 5$ GHC, then a three-digit identifier from 000 to 234 will work. Based on the rules of how the structured topologies are built, we can always design an identifier system which we can refer to find their neighbors, thus result in convenience in routing. So, the inter-identifier identifies the random cluster each node resides in and eases the routing scheme's design at the level of structured topology.

The inner-identifier is the most challenging part since boundary nodes are chosen randomly. In the edge coloring theory, whether a graph is class 1 is a typical NP-complete problem, which cannot be solved in polynomial time. We employ DSATUR [10], the best-known heuristic algorithm in this area, to approximate the optimal solution. Such an algorithm usually generates multiple coloring strategies, one of which will be randomly selected. Basically, each color identifies a specific inner-identifier. As shown in Fig. 5.2, the structured links are colored with four colors, i.e., black, purple, orange and blackish green, which represent inner-identifier 000, 011, 010 and 001, respectively. Furthermore, the inner-identifier of each boundary node in a random cluster is assigned with the identifier of the associated structured link. As shown in Fig. 5.2, all boundary nodes are assigned the inner-identifier with 000, 011, 010 and 001, respectively. As for the rest nodes in the random cluster, we calculate the inner-identifier interval according to the binary system. For example, if there are 9 nodes in a random cluster, a 4-digit binary range from 0000 to 1000 will identify all of them. Then the identifiers in the interval, except those used as boundary nodes, are assigned to the non-boundary nodes randomly.

5.3.2 Identifier-Based Routing Algorithm

According to allocated identifiers of all nodes, especially those boundary nodes, we derive an efficient routing algorithm for R3. Generally, the transmission of data flows between any pair of nodes can be usually divided into a series of inter-cluster and intra-cluster routing. Such two kinds of routing are totally different because of the lack of structured links inside each random cluster.

The intra-cluster routing is just the same as routing in a random graph like Jellyfish or Scafida. But in our cases, routing can be simpler since the number of nodes in our random cluster is much less than that of Jellyfish or Scafida. Typically, we employ the shortest k-path Algorithm to search the paths between any pair of nodes. Furthermore, ECMP protocol can be used to control data transmission and avoid congestion.

The inter-cluster routing aims at finding the shortest path from the source to destination on the random cluster level. To be specific, we need to determine the relay random clusters and all boundary nodes on the way. Inter-cluster routing consists of two steps. First, calculating the relay random clusters from the source cluster, where the source node locates, to the destination cluster, where the destination node locates. Since the inter-identifiers contain the topology information of the structured graph, which the hybrid topology utilizes, the relay random clusters can be easily derived from the inter-identifiers of the source node and destination node. Second, determining the boundary nodes of all related random clusters that have been derived from the first step. According to the colors of structured links we have allocated, the inner-identifiers of each boundary node along the random cluster level path will be gained. This work is straightforward since the inner-identifiers of the two endpoints, which connect with the same structured link, share the same identifier with the colored link. In this special way, the random cluster level path can be derived based on the structured link colors and the identifier system we have established before.

From the global viewpoint, given a pair of the source node and destination node, first of all, the routing algorithm judges whether they belong to the same random cluster according to their inter-identifiers. If yes, then the intra-cluster routing algorithm will be employed to find the path. On the contrary, then inter-cluster routing will provide the random cluster level path; hence, all relay random clusters and relay boundary nodes are determined. Then, the path needs to be specified at a finer level. That is, each relay random cluster along the random cluster level path will employ intra-cluster routing to find the relay nodes inside them. With the relay nodes and links inside each relay random cluster and the structured links added into the path, the whole routing process is accomplished.

Algorithm 2 Routing in hybrid topologies

Require: The Hybrid topology, H; The source node src, and its identifier $iden$-src; The destination node dst, and its identifier $iden$-dst; The path number k; the number of digit in inter-identifier x.

1: Coloring the links and allocate identifiers;
2: Let tem be a integer with default value 0;
3: Let $iden1$ and $iden2$ be a identifier respectively;
4: **if** $GetInterIden(iden$-$src,x)$ == $GetInterIden(iden$-$dst,x)$ **then**
5:     $path$ = kStar($iden$-$src,iden$-dst,k);
6: **else**
7:     Get the inter-identifier of clusters, denoted as $inter$-$iden$;
8:     Get structured links needed, denoted as $structured$;
9:     Get color of links needed;
10:    Get inner-identifier of boundary nodes, denote as $iden$-$color$;
11:    **while** $tem < iden$-$color.size$ **do**
12:        $iden1 \longleftarrow iden$-$src$;
13:        $iden2 \longleftarrow inter$-$iden[tem]+iden$-$color[tem]$;
14:        $path$ += $Dijkstra(iden1,iden2)$;
15:        $path$ += $inter$-$iden[tem])$;
16:        tem++;
17:    $path$ += $Dijkstra(iden2, dst)$;
18: **return** The routing path $path$.
19: **function** GETINTERIDEN(iden, x)
20:    **for** $i = 1$ to x **do**
21:        $inter$-$iden$ += $coor[i]$;
       return $inter$-$iden$.

As explained in Algorithm 5.2, given two nodes, we first judge whether they belong to the same random cluster according to their inter-identifiers. If it is true, K^* Algorithm [11], the most effective heuristic search algorithm so far, is adopted to search k shortest routing paths. If they reside in different random clusters, Algorithm 5.2 identifies the structured links and their colors and find those boundary nodes in each relay cluster. We then add links to the path iteratively by invoking the Dijkstra algorithm in each cluster. In Fig. 5.2, a source node with the red color in cluster 00 needs to communicate with a destination node with the yellow color in cluster 13. Their identifier is given as 00101 and 13100, respectively. Obviously, they belong to different random clusters, and there exist two routing paths between clusters 00 and 13, i.e., $00 \rightarrow 03 \rightarrow 13$ and $00 \rightarrow 10 \rightarrow 13$. We use the first path as an example, 03 is a relay cluster on the random cluster level. The purple and orange links associated with inner-identifiers 011 and 010 are two inter-cluster links. Furthermore, the boundary nodes can be located, i.e., 00001 in cluster 00, 03011 and 03010 in cluster 03, and 13010 in 13 clusters. Then, the Dijkstra algorithm is utilized to derive a path of added links inside each random cluster, e.g., the shortest path from 00101 to 00011 in cluster 00, from 03011 to 3010 in cluster 03, and from 13010 to 13100 in cluster 13. In this way, any pair of nodes can finally achieve the routing path inside R3.

When nodes are added or eliminated from R3, the routing tables need to be updated. Note that, unlike other topologies where the addition or deletion of a node may affect the global routing, R3 suffers the least since it limits the influence of topology alteration into the specific random cluster. Here is an example, for Fat-Tree, if one of the core switches breakdowns, then the routing tables of all nodes that belong to the subtree rooted from the failed switch will be updated to suit the new topology. On the contrary, when 00001 fails in Fig. 5.2, only those nodes of the 00 cluster may need to update the routing table, thus will never impact nodes in other random clusters.

5.4 Topology Optimization

Given the number of servers in a data center, a typical question is how to allocate the ports of each switch when establishing the data center networking structure. That is, how many ports should each switch be allocated to connect with servers? Note that the remained ports of each switch are utilized to form a networking structure among all switches. There lies a trade-off between the number of switches and the network diameter since the increasing number of switches leads to decreased network diameter at the cost of incurring more investment [12]. Meanwhile, the ratio of node degree to the network diameter is a classical problem for topology design. In the design of our hybrid topologies, more structured links can ease the routing. On the contrary, networking can benefit from more randomness since random topologies can naturally support the incremental expansion with a low diameter. In this case, how much randomness is optimal for both routing and networking? To answer these questions, the proposed hybrid topologies, i.e., R3, need further optimization.

5.4.1 Impact Factors of the R3 Topology

Given the number of servers and that of ports at each switch, how many ports should each switch be allocated to connect with servers and other switches, respectively, so as to realize the minimum APL with an acceptable amount of switches? In R3, to minimize average path length, we must concern at least three impact factors. As a hybrid topology combine r-RRGs and GHC via the compound graph theory, the diameter of R3 is decided by three factors. That is 1) the dimension of GHC, $m_s, m_{s-1}, \ldots, m_1$, 2) the node degree of RRGs, r, and 3) the number of nodes in each RRG, $n_1, n_2, \ldots, n_t$, where $t = m_s \times m_{s-1} \times \cdots \times m_1$. The dimension influences the redundancy of the routing path and the number of structured links in routing paths. Moreover, it dominates the Hamming distance between any pair of nodes, x and y, in terms of the number of hops between them [13]. The other two factors determine the number of relay nodes inside each cluster along the path. APL is a network-level measurement, so we calculate it and integrate such factors to reveal their influence on the network diameter.

5.4.2 Optimization Strategy of Topologies

In R3, each routing path consists of two parts, the structured links and some random link in random clusters on the path.

Theorem 5.3 *In R3, let APL_{ghc}, APL_{rrg}, and APL_{r3} denote the average length of routing in GHC, RRGs and R3, respectively. Then we have*

$$APL_{r3} = APL_{ghc} + (APL_{ghc} + 1) \times APL_{rrg}.$$

Proof In such a routing path, there exist APL_{ghc} structured links, and $APL_{ghc} + 1$ random clusters among which $APL_{ghc} * 1$ random clusters are relay. In each of such clusters, the path length is APL_{rrg} on average. Thus, Theorem 5.3 is proved.   □

Theorem 5.4 *For a $G(m_s, m_{s*1}, \ldots, m_1)$, $N = m_s \times m_{s-1} \times \cdots \times m_1$ denotes the number of nodes, x_l denotes the number of node pairs, each of which exhibits the distance of l. We then have*

$$x_l = (N/2) \sum_{i_1=1}^{s-l+1} \cdots \sum_{i_j=j}^{s-l+j} \cdots \sum_{i_l=l}^{s} [(m_{i_1} - 1) \dots (m_{i_l} - 1)] \tag{5.2}$$

$$APL_{ghc} = \left(2 \sum_{l=1}^{s} l \times x_l \right) / (N(N-1)) \tag{5.3}$$

Proof Consider a node A, denoted as $y_s y_{s-1} \dots y_1$, in this GHC, If another node B is l hops away from node A, the coordinates just differ in l dimensions Thus, APL_{ghc} can be calculated naturally based on x_l. Thus Theorem 5.4 is proved. $\qquad \square$

Note that different structures exhibit varied APL when the node degree and number of nodes do not change [7]. The default value of APL in this chapter is set to $\log_{r-1} n$, where n and r denote the number of nodes and the degree of each node, respectively. Theorems 5.3 and 5.4 derive a model of the network diameter for R3. We consider a special case that $n_1 = n_2 = \cdots = n_t = n$. We build this optimal model with the total server number T, the port count of each switch p as input. α and β denote the number of ports each switch allocates to connect with other switches and servers, respectively. Let $t \in [1, T]$ be the total number of used switches. The topology optimization can be modeled as follows:

$$\min \ APL_{r3} \quad s.t. \quad \begin{cases} T \le s \times \beta \\ t \le n \times \prod m_i \\ \sum_{i=1}^{s} (m_i - 1) + 1 \le n \\ 2 \le \alpha + \beta \le p \\ 1 \le r \le \alpha - 1 \end{cases} \tag{5.4}$$

In this model, the minimum APL is searched in the domain of feasible solution with five constraints satisfied. The first inequality promises that the number of servers that our topology can accommodate is not less than the input T. The second inequality constraints the number of switches the topology can accommodate is more than the number of switches that are actually used. The third inequality guarantees that our routing algorithm can be built successfully. While, the last constraint reveals that each switch should leave at least one port for structured links. In fact, this model is a nonlinear integer programming problem, which is NP-hard. According to the optimal gradient algorithm, we utilize an associated tool (ModelCenter) to search the minimum result.

Figure 5.3a shows that the optimized APL of R3 increases when T varies from 1000 to 6000. This result is reasonable because more servers need to be accommodated. It, however, fluctuates at 4500 and 5500 because there are more switches and fewer ports link with servers in these two cases. For this reason, the topology exhibits lower APL at the cost of increasing the investment due to more switches. The optimal gradient algorithm searches towards the fastest decline direction and finds the minimal value. Figure 5.3b indicates that, given $T = 2000$ and $p = 48$, the search process

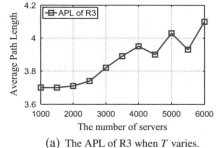

(a) The APL of R3 when T varies.

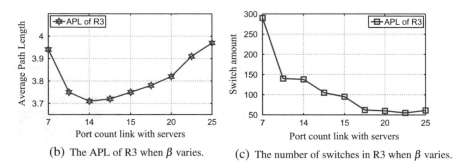

(b) The APL of R3 when β varies.   (c) The number of switches in R3 when β varies.

Fig. 5.3 The APL of R3 with different parameter settings

terminates when $\beta = 15$, and the minimum APL is 3.71. Furthermore, we evaluate the number of switches after APL is optimized. Figure 5.3b and Fig. 5.3c demonstrate the marginal effect around the extreme point. When $s = 62$, $APL = 3.78$ is a little bit higher than 3.71. However, 138 switches are required so as to reach the extreme point. This will double the investment and obviously not be cost-effective. So, whether the minimum APL is the best choice depends on designers.

5.5   Incremental Expansion of R3

Incremental expansion is essential to data centers since they are usually required to accommodate the arbitrary number of servers on demand. For data centers based on our hybrid topologies, two methods can be employed to realize the incremental expansion, i.e., expansion by adding nodes in an existing random cluster and expansion by inserting a new random cluster.

5.5.1 *Expansion within an Existing Random Cluster*

Random topologies like RRG and scale-free networks support incremental expansion naturally. New nodes can be added one by one. For RRG, when a new node is added in, several existing nodes break up their links and connect to the new one [5]. For a scale-free network, according to its generation algorithm, a new node will be linked with m existing nodes, which are selected based on the preferential attachment principle [6]. To keep from the unbalance and bottleneck, those random clusters with the minimum number of nodes are chosen to host new nodes. With more and more nodes are added into those selected clusters, the length of the inner-identifier needs to be increased to maintain consistency in the whole network. In other words, the length of in inner-identifier is decided by the maximum number of nodes in random clusters. In Fig. 5.2, the number of nodes in each cluster is no more than 8. A 3-digit inner-identifier is feasible. Once a new node is added into cluster 00, the maximum number of nodes is 9. Thus, the existing 3-digit inner-identifier is replaced by a 4-digit inner-identifier. That is, existing inner-identifiers are updated by adding a new digit in the front of them. Theoretically, the number of the node that each random cluster can accommodate is unlimited. However, if the random clusters have too many nodes, the structured links may be a bandwidth bottleneck since these random clusters are interconnected with structured links only.

5.5.2 *Expansion by Adding an Extra Random Cluster*

As mentioned above, for the sake of network performance, it is not allowed to add too many nodes to existing random clusters. When all random clusters can no longer accommodate more nodes, we consider adding new random clusters to achieve the scale expansion of R3. However, many structured topologies generally adopt a hierarchical or recursive approach to achieve scale expansion, which obviously can not guarantee incremental expansion. Therefore, this chapter proposes a new expansion method for R3, which enables its structured topology to expand incrementally like a hypercube structure.

Definition 5.4 In a n-dimensional hypercube, a $(n - 1)$-dimensional coordinate system can be established to identify each node according to its building rule. We call a pair of nodes corresponding if their coordinates are different only at the first digit.

As shown in Fig. 5.4b, a new cluster 100 and an existing cluster 000 are called a pair of corresponding clusters. Algorithm 5.3 depicts a framework of cluster-level expansion. First, the Algorithm judges whether the existing inter-identifiers are sufficient to distinguish all random clusters. If not, the Algorithm adds one digit to identifiers and updates the identifiers of existing clusters. Second, random clusters are added according to the following steps.

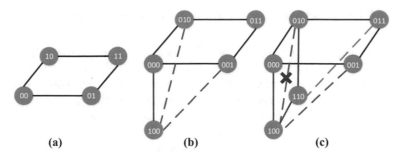

Fig. 5.4 A example of expansion by adding extra clusters

1. Assign inter-identifiers to new random clusters.
2. Find existing clusters whose Hamming distance is one from the new cluster based on the expansion rules of the hypercube, and connect those clusters with the new cluster.
3. Delete unnecessary links.
4. The lack of some neighbor nodes makes the number of clusters insufficient to construct a complete hypercube. To solve this problem, it needs to connect the new cluster with existing clusters. Specifically, connect the new cluster with the neighbors of its corresponding cluster.

We depict an example in Fig. 5.4 when a new cluster wants to add into a quadrangle, the coordinate of existing clusters is increased one digit, and the new cluster is identified as 100. According to the quadrangle rule, cluster 100 should connect with cluster 000. Then the degree of cluster 100 is only 1, while the required degree is 3. As shown in Fig. 5.4b, the Algorithm finds cluster 000, which is the corresponding cluster of 100 and connects cluster 100 with the neighbors of cluster 000, i.e., clusters 010 and 001. As depicted in Fig. 5.4c, when cluster 110 is added to the existing network, the algorithm will connect 110 and 100, while deleting the link between 100 and 010 which is added before to maintain the characteristics of the hypercube.

Algorithm 3 Expansion with Extra Random Clusters, G

Require: The regular topology, G; The amount of new random clusters, θ.
1: Let $New[\theta]$ be those θ new clusters that will be added;
2: **for** $i = 0$ to $\theta - 1$ **do**
3:     Assign inter-identifier to $New[i]$;
4:     Find clusters whose hamming distance are 1 from $New[i]$, and connect those clusters with $New[i]$;
5:     Delete unnecessary links;
6:     **if** the number of $New[i]$'s neighbors are less than expected **then**
7:         Find corresponding node of $New[i]$;
8:         Connect $New[i]$ with neighbors of its corresponding node;
9: **return** The new structured topology G.

5.6   Performance Evaluation

In this section, we simulate the hybrid topology R3 to evaluate the routing flexibility, cabling cost and network performance. Typically, we compare R3 with a fully structured GHC and a fully random Jellyfish topology, respectively.

5.6.1   The Routing Flexibility

The building routing table in a large-scale data center is tough work for those fully or partially random DCNs, due to the huge number of links and potential paths between any pair of switches. To measure the time consumption due to finding the shortest path between each switch pair, we conduct a series of experiments with different settings of switches.

We construct R3 with different amounts of switches, ranging from 200 to 900. In each of these R3s, 20 switches form an 8-RRG random cluster, and such random clusters are connected via structured links. Since our routing method is coloring-based, the time consumption of routing consists of two parts, i.e., the coloring time and the routing time. During the coloring period, each structured link is colored with one color to record the nodes used to link each pair of random clusters.

As demonstrated in Fig. 5.5a, compared with the coloring time, the routing time contributes most of the total time. Figure 5.5b reports that our routing method outperforms the traditional Dijkstra algorithm in R3. More precisely, the coloring-based routing algorithm reduces half of the time consumption compared to the Dijkstra algorithm on average. The reason is that, compared with the traditional Dijkstra algorithm, which is time-consuming, the coloring-based routing algorithm limits the Dijkstra algorithm inside each random cluster only. Furthermore, as reported in Fig. 5.5c, R3 consumes much less time compared to the Jellyfish topology. Note that the building clusters of our hybrid R3 topology are 8-RRGs. After introducing structured links into the topology, the total number of links in R3 is between that of 8-RRG Jellyfish and that of 9-RRG Jellyfish. The evaluation result, however, demonstrates that the routing time of R3 is less than that of 8-RRG and 9-RRG Jellyfish many times.

Additionally, routing flexibility is important in DCNs, where failures of commodity devices are very common. In R3, once a switch breaks down, the coloring-based routing algorithm will immediately derive another available path and update the involved routing table dynamically.

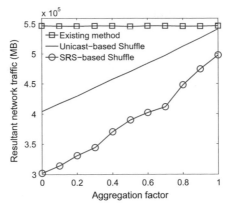

(a) Time consumption of routing and coloring.

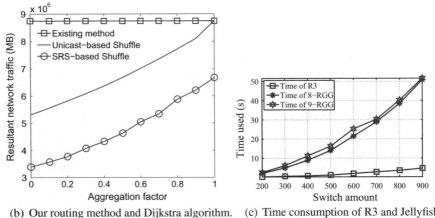

(b) Our routing method and Dijkstra algorithm. (c) Time consumption of R3 and Jellyfish.

Fig. 5.5 Time consumption of routing process

5.6.2 The Cabling Cost

In DCNs, a huge number of links are utilized to interconnect large-scale switches and servers to form a designed topology. For each of our hybrid topologies, those long-distance random links will significantly increase the cabling cost and complexity. In this section, we calculate the total length of all cables in R3 and Jellyfish to evaluate the cabling cost.

Racks in a data center are placed as a matrix for cooling and maintaining purposes. To minimize the total length of all cables, we suppose that racks are placed as a quadrate or quadrate-like structure. Given the number of switches, we first calculate the length and width according to the quadrate-like location strategy. Then, we calculate the total length of all links via the Pythagorean Theorem. Meanwhile, we assume that links between racks are underground distribution such that the geographical distance can be calculated as the link distance.

Fig. 5.6 Cabling cost
comparison with Jellyfish

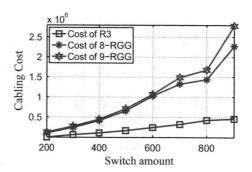

Reasonably, we use the distance between any pair of racks as the metric of cabling cost. Figure 5.6 depicts the cabling cost of R3 compared to that of Jellyfish. For fairness, we compare R3 with 8-RRG Jellyfish and 9-RRG based Jellyfish, respectively. R3 causes much less cabling cost than 8-RRG as well as 9-RRG based Jellyfish. This result is reasonable since more remote links are introduced into Jellyfish with the increasing number of switches. However, the distance between any pair of switches in R3 is predictable and less than that in Jellyfish.

5.6.3 The Network Performance

We compare R3 with the generalized hypercube and Jellyfish, which represent fully structured and random topologies, respectively. To evaluate their performance, we vary the amount of 24-ports switches from $n = 8$ to 360. We calculate the maximal amount of servers they can accommodate, i.e., network order, and evaluate the throughput under all-to-all traffic. In the best case, each switch inside a random cluster has a structured edge to connect with another switch in other clusters. So, R3 has the same network order as RRG. But for GHCs, the network order depends on its dimensions. Consider $n = 200$ as an example and there is $200 = 2 \times 2 \times 2 \times 5 \times 5$. Each switch should reserve at least 11 ports for linking with other switches, and the remainder 13 ports link with servers. Consequently, the network order is at most $200 \times 13 = 2600$. To construct the densest R3, we use 5×5 GHC and 7-RRG, which lead to a hybrid topology, which can accommodate $16 \times 200 = 3200$ servers, and the maximal node degree is 8. For fairness, an 8-degree Jellyfish is built as a reference. Figure 5.7 depicts the resulting network order. It is clear that both R3 and Jellyfish can accommodate a large number of servers. This is reasonable because both R3 and Jellyfish have larger design space than GHC.

Indeed, the throughput of DCNs is affected by not only the topology but also the bandwidth allocation strategy [14, 15]. For a given DCN, different bandwidth allocation strategies result in different network performances. In this chapter, to comprehensively reveal the impact of topologies, our experiments focus on the topologies

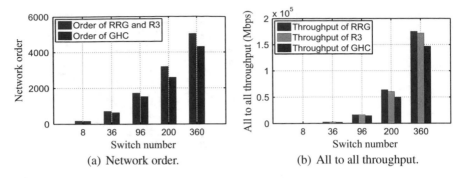

Fig. 5.7 Performance comparison between R3, GHC and Jellyfish

under the same bandwidth allocation strategy. We first verify the network performance of R3 under different amounts of switches and then compare the network
throughput with Jellyfish and GHC under all-to-all traffics. Typically, the bandwidth
of each link is set to be 1000 Mbps, and the data rate of each server is 100 kbps. We
monitor each flow via the flow monitor function in NS3 and obtain the total throughput by summing up the data rate of each flow. Figure 5.7b plots that the network
throughput of R3 is always a little bit less than that of Jellyfish but much more than
that of GHC. Thus R3 integrates the advantages of both GHC and Jellyfish while
abandons their weakness.

5.7   Discussion

The proposed hybrid topology R3 combines structured topologies with random
topologies seamlessly via the compound graph. To fully understand the hybrid
designing methodology, we discuss the following issues further.

Rethinking the routing algorithm. The proposed edge-coloring based routing
algorithm achieves fast and accurate routing tables. In effect, such a routing algorithm
is propagable since Theorem 5.1 guarantees that at most $\Delta + 1$ kind of colors are sufficient to color all edges. Namely, the novel routing algorithm works well even though
R3 selects different structured topology. Let m denotes the average number of nodes
in random clusters, $|E|$ and $|V|$ denote the number of edges and vertex in the chosen
structured topology, respectively. Then the time complexity of our routing algorithm
is $O(|E| \times |V|) + O(|V| \times m^2)$, where $O(|E| \times |V|)$ is the time complexity of col-
oring and $O(|V| \times m^2)$ is the time complexity of routing inner random clusters.

Dedicated integer programming model. Note that the proposed integer programming model is used to find the best parameter setting of a given structured
topology, the number of servers and amount of ports per switch. So, different structured topologies result in different integer programming models to describe the hybrid
topologies precisely.

Incremental expansion of our hybrid topologies. As for the expansion issues for hybrid topologies, we propose two appropriate methods, i.e., the expansion within an existing random cluster and the expansion by adding an extra random cluster. The newly added servers ought to be allocated into the existing random clusters evenly other than embedding them into one or several clusters in bathes. Besides, which expansion strategy should be employed remains an open problem for the designers so that the flexibility and design space will be guaranteed.

The experiment methodology. Our experiments concentrate on evaluating the performance of R3, RRG and GHC since R3 is constructed via combining RRG and GHC seamlessly. Different selections of structured and random topologies will definitely result in different performances. The performance of each of the other hybrid topologies is similar with R3, i.e., its performance falls between that of the used structured topology and that of the utilized random topology. Thus, the major motivation of our experiment is to prove that the proposed hybrid designing methodology combines the benefits of both structured and random topologies together while avoiding their weakness successfully.

Constraints of hybrid DCNs. The proposed hybrid DCNs face a few potential limitations. First of all, three constraints must be satisfied to achieve a hybrid topology, as mentioned before. Second, if there exist too many nodes in a random cluster, the boundary nodes of that cluster may be overloaded and thus may result in congestion. In other words, the designer must concern the capacity of each boundary node when building or upgrading the hybrid data center network.

References

1. Al-Fares M, Loukissas A, Vahdat A. A scalable, commodity data center network architecture [J]. ACM SIGCOMM Computer Communication Review, 2008, 38(4): 63–74.
2. Greenberg A, Hamilton J R, Jain N, et al. VL2: a scalable and flexible data center network [J]. ACM SIGCOMM computer communication review. ACM, 2009, 39(4): 51–62.
3. Guo D, Chen T, Li D, et al. Expandable and cost-effective network structures for data centers using dual-port servers [J]. Computers, IEEE Transactions on, 2013, 62(7): 1303–1317.
4. Shin J Y, Wong B, Sirer E G. Small-world datacenters [C]. In Proc. of 2nd ACM SOCC, Cascais, Portugal, 2011: 1–13.
5. Singla A, Hong C Y, Popa L, et al. Jellyfish: Networking Data Centers Randomly [C]. In Proc. of 9th USENIX NSDI, Boston, USA, 2011: 17–17.
6. Gyarmati L, Trinh T A. Scafida: A scale-free network inspired data center architecture [J]. ACM SIGCOMM Computer Communication Review, 2010, 40(5): 4–12.
7. Bollobs, B. Random Graphs[M]. Cambridge: Cambridge University Press, 2001.
8. Xie J, Guo D, Xu J, et al. Efficient Multicast Routing on BCube-Based Data Centers [J]. KSII Transactions on Internet and Information Systems (TIIS), 2014, 8(12): 4343–4355.
9. Bondy J A, Murty U S R. Graph theory with applications [M]. London: Macmillan, 1976.
10. Brlaz D. New methods to color the vertices of a graph [J]. Communications of the ACM, 1979, 22(4): 251–256.
11. Aljazzar H, Leue S. K*: A heuristic search algorithm for finding the k shortest paths [J]. Artificial Intelligence, 2011, 175(18): 2129–2154.

12. Giannini E, Botta F, Borro P, et al. Platelet count/spleen diameter ratio: proposal and validation of a non-invasive parameter to predict the presence of oesophageal varices in patients with liver cirrhosis [J]. Gut, 2003, 52(8): 1200–1205.
13. Bhuyan L N, Agrawal D P. Generalized hypercube and hyperbus structures for a computer network [J]. Computers, IEEE Transactions on, 1984, 100(4): 323–333.
14. Guo J, Liu F, Zeng D, et al. A cooperative game based allocation for sharing data center networks [C]. In Proc. of IEEE INFOCOM, Turin, Italy, 2013: 2139–2147.
15. Guo J, Liu F, Huang X, et al. On efficient bandwidth allocation for traffic variability in data-centers [C]. In Proc. of IEEE INFOCOM, Toronto, Canada, 2014: 1572–1580.

Chapter 6
VLCcube: A Network Topology for VLC Enabled Wireless Data Centers

Abstract Recent results have made a promising case for offering oversubscribed wired data center networks (DCN) with extreme costs. Inter-rack wireless networks are drawing intensive attention to augment such wired DCNs with a few wireless links. For this reason, we propose a hybrid network topology, VLCcube, which consists of wireless links and wired links, by introducing extra visible light communication (VLC) links into data centers. Specifically, VLCcube augments Fat-Tree, a representative DCN in production data centers, by installing 4 VLC transceivers on top of each rack for constructing four wireless links of 10 Gbps. These wireless links improve the network capacity of wired links and organize all racks into a wireless Torus structure. This chapter focuses on the topology design, hybrid routing, and flow scheduling schemes for VLCcube, and carries out related experiments to show its performance. VLCcube has better network performance than Fat-Tree since many flows transmitting along 4-hop paths in Fat-Tree can achieve a shorter transmission range if they are transmitted in the wireless Torus. The congestion-aware flow scheduling algorithm can further improve the performance of VLCcube. Actually, VLCcube is just one alternative to the utilization of VLC links in data centers. The introduction of VLC links integrates seamlessly with the existing DCNs, and it improves the performance and design flexibility of DCNs effectively.

6.1 Introduction

6.1.1 Motivation

Data centers have emerged as infrastructures for online applications and infrastructural services. Thousands of servers and switches are interconnected via a specific data center network (DCN). DCNs can be roughly divided into two categories. The first category is wired DCNs, each of which connects all switches and servers with wired links via cables, fibers or twisted-pair links. Fat-Tree [1] and VL2 [2] fall into this category. The second one is wireless DCNs, which employ wireless links to argue a wired DCN or organize servers and switches as a fully wireless network topology [3, 4].

© Springer Nature Singapore Pte Ltd. 2022
D. Guo, *Data Center Networking*,
https://doi.org/10.1007/978-981-16-9368-7_6

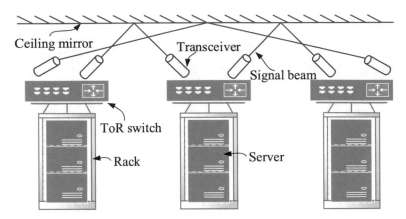

Fig. 6.1 An example of wireless DCN, in which the racks are interconnected with wireless links

Wired DCNs are widely adopted. However, they suffer from inherent challenges: (1) They are either overprovisioned with good performance but the high cost or oversubscribed with low cost but poor performance. (2) It is extremely costly and complicated when expanding a wired data center. (3) They cause vast cabling and maintenance cost [5]. (4) Large-scale wired DCNs usually adopt multiple-level structures. As a result, two servers, which across racks, must employ the upper-level links to communicate with each other, even if they are very close physically.

To eliminate the non-trivial cost and increase the flexibility during the expanding process of any wired production DCN, several wireless DCNs are proposed at the inter-rack level. As depicted in Fig. 6.1, the racks are connected with the introduced wireless links. Typically, the radio frequency (60 GHz) [4] and free-space-optical (FSO) communication techniques [3] are employed to establish an inter-rack wireless network. These proposals can considerably improve the performance of any existing wired DCNs in terms of bandwidth and packet latency [6]. Moreover, the wireless links can be dynamically reconfigured to meet the demand of the undergoing flows.

Inspired by the promise of easy-deployable and plug-and-play, we envision a radically different design of the inter-rack wireless network, which should simultaneously concern the following three design rationales: (1) all inter-rack links are wireless; (2) without imposing any infrastructure-level alteration on the existing wired production data centers; and (3) the inter-rack wireless network is plug-and-play and has no need of additional mechanical or electronic control operations.

This vision, if realized, will lead to unprecedented benefits for wireless DCNs. Firstly, it ensures high flexibility and low cabling cost of the network by introducing wireless links on demand. Secondly, it simplifies the configuration and usage process of the inter-rack wireless network due to bringing no additional mechanical or electronic control operations. Such simplification makes the inter-rack wireless network extremely compatible with existing wired DCNs. Thirdly, it alleviates the burden of managing and maintaining a data center. Once those wireless links are established, they will work permanently without additional control operations.

Existing proposals on the inter-rack wireless network focus on the flexible recon-figuration of links. Such proposals, however, do not consider the other two essential design rationales. First of all, they have to update or even reconstruct the deploy-ment environment of existing production data centers. For example, prior proposals using 60GHz, as well as FSO communication, have to decorate the ceiling to be a huge mirror to achieve over-the-horizon communication. Besides, to realize flex-ible reconfiguration, dedicated optical devices are required, e.g., ceiling mirrors, plano/biconvex lens. Moreover, they impose frequent and complicated control on wireless devices and peripheral equipment when configuring a wireless link.

In this chapter, we propose VLCcube to achieve the above three design rationales simultaneously. VLCcube augments Fat-Tree, a representative production wired DCN, by organizing all racks into a wireless Torus structure via the emerging visi-ble light communication (VLC) techniques. Hence, it is a hybrid network topology for data centers, by seamlessly integrating the wired Fat-Tree and wireless Torus. Although the 60GHz and FSO communication techniques are also suitable for VLC-cube in theory, we prefer the VLC since it is becoming a promising choice for the next-generation wireless technology by offering low cost, unregulated bandwidth and ubiquitous infrastructures support. Inherently, VLC links eliminate the peripheral devices except for VLC transceivers and need no additional mechanical or electronic control.

6.1.2 Related Work

Due to the essential status of DCNs, a huge body of work has been conducted to improve network performance. The representative DCNs can be classified into two categories, i.e., the wired DCNs and the wireless DCNs.

Typically, the wired DCNs take advantage of the merit of excellent topologies, e.g., Torus, Hypercube, Kautz, Small-world, etc. Fat-Tree, VL2 and Portland all adopt the multi-level structure Fold-Clos, which guarantees high fault-tolerance and high network performance. The server-centric DCNs, such as BCube [7], and DCell [8], allows servers to participate in the network routing and forwarding. Besides, some random structures are introduced into DCNs for the purpose of incremental expansion, such as Jellyfish [9] and Small-world [10]. However, it is hard to balance the cost and network performance in wired DCNs. On the one hand, some data centers are over-provisioned to accommodate the potential peak traffic, leading to high hardware costs and low utilization rates. On the other hand, some wired data centers are over-subscribed with low cost, but poor performance [3].

To solve the above problems, many wireless communication technologies are introduced into data centers to improve their network performance and structure flexibility [5]. Among them, 60GHz radio frequency communication technology is one of the first to be used in data centers [4, 11]. It requires altering the ceiling to huge reflectors to support long-range communication. When constructing wireless links between any pair of racks, the source rack must accurately adjust the transceiver angle to ensure that the 60GHz signals reflected by the ceiling reflector can be received by the transceiver of the destination rack. Similarly, Firefly [3] adopts FSO to construct the wireless network, and it dynamically reconfigures the wireless network according to the traffic demand. Note that the dynamical reconfiguration of the wireless network depends on the complicated mechanical and electrical control of optical devices.

In this chapter, we propose a novel wireless network topology at the rack level, which achieves the three design rationales. For this purpose, we organize the existing wired DCN into a wireless Torus using VLC links in the rack level. In a two-Dimensional Torus, each rack naturally connects with at most four adjacent racks. VLCcube does not require the ceiling reflector to reflect and relay wireless VLC signals. Adjacent racks are connected through permanent VLC links after transceivers are configured. These wireless links work without the need to manipulate other devices to cooperate with the use.

6.2   The Design of VLCcube

We first discuss the feasibility and interference issue of interconnecting racks using VLC links and accordingly design a novel VLCcube topology. It seamlessly augments the wired data center Fat-Tree, using a wireless inter-rack Torus network.

6.2.1   Feasibility of Introducing VLC Links into DCNs

For VLC, transmitting data is achieved by intensity modulation of visible spectrum lighting emitting diodes (LEDs) or laser diodes (LDs). On-Off keying modulation scheme, where "ones" and "zeros" are represented by the presence or absence of light, is the simplest form of digital communication [12]. To employ the VLC links, three vital issues have to be concerned, including the data rate, transmission distance, and the accessibility of devices.

Data rate. By employing those high switching frequency LEDs, a single color VLC link can realize a considerably high data rate up to 3 Gbps [13]. Such devices could potentially deliver data rates in the order of 10 Gbps by using RGB triplet. Besides, a single laser beam can even achieve 9 Gbps data rate by employing the 450-nm GaN LDs [14]. Hence, we believe that the data rate of VLC links is capable of transmission in data centers.

Transmission distance. The LED-based VLC links can achieve about 10 Gbps data rate within 10 meters, which is sufficient to interconnect two close racks inside a data center. We notice that a project named Rojia prolongs the distance of VLC to 1.4 kilometers, but with a limited data rate [15]. Besides, the LD-based VLC links can realize fast long-distance communication (in the order of kilometers) with high data rate [16], due to its outstanding directionality. Hence, the LED-based VLC links can be employed as the short links, while the LD-based VLC links are competent to the long-distance transmission in DCNs.

Accessibility. The off-the-shelf full-duplex VLC devices, i.e., transceivers, are developed and released [15, 17]. A development platform called MOMO has delivered API and SDK for users to customize their VLC-based applications. For example, the VLC techniques have seamlessly integrated into a platform of the internet of things. Moreover, pureLiFi [17] provides the opportunity for customers to rapidly develop and test VLC applications for cost-effective, high-speed data communication solutions by using commercial LED infrastructures.

Accordingly, it is reasonable to employ the VLC links to augment wired DCNs, without incurring additional cabling costs or modifying the hosting environment of data centers.

6.2.2 The Interference Among Transceivers

The benefits of VLC links motivate us to employ them to augment the wired inter-rack networking in data centers. However, interference is an essential obstacle when utilizing VLC links.

Typically, on the top of each rack, a few VLC transceivers should be set up to organize the ToR switches as a dedicated wireless topology. Given a rack R, when multiple neighboring racks send data to it simultaneously, interference occurs if multiple transceivers on R can perceive the light from different source racks but fail to distinguish them.

To evaluate the VLC interference, we conduct simulations using professional optical software, i.e., TracePro70 [18]. We place four receivers with orthogonal orientations on the top of a rack, which are denoted as T_1, T_2, T_3 and T_4, respectively. Then, a batch of visible light is emitted towards T_1. The irradiance map of each receiver can identify how much light has been detected by other receivers. If T_2, T_3 and T_4 detect intensive light, it demonstrates that the interferences to them are prominent.

Figure 6.2a depicts the result observed by T_1. Obviously, the receiver detects the majority of the emitted light, and the central part of the receiver captures most of them. Due to the scattering, some light deviates from the central line; hence, the non-central areas can also detect light. By contrast, as shown in Fig. 6.2b–d, the other three receivers can hardly capture the light since the normalized irradiance is only 0.001 in several points in such figures. We also note that, Fig. 6.2c records the least irradiance at T_3. Consider that T_3 is right behind T_1 and it is difficult for the light to pass by T_1 to reach T_3. Consequently, the light towards T_1 results in limited interference to the other three receivers. This observation shows that deploying four transceivers on the top of a rack is feasible and will bring negligible interference. Accordingly, we will design the wireless topology of the VLCcube, where each rack owns just four VLC transceivers.

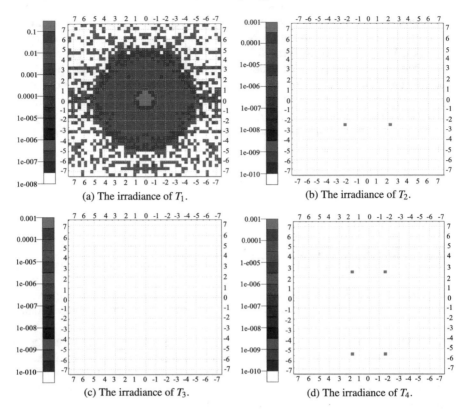

(a) The irradiance of T_1.

(b) The irradiance of T_2.

(c) The irradiance of T_3.

(d) The irradiance of T_4.

Fig. 6.2 The irradiance of each receiver in the simulation

6.2.3   Topology Design of VLCcube

Inside a data center, each server connects to the ToR switch inside a rack. All racks usually form a hierarchical network topology by using additional upper-level wired links and network devices rather than connect with each other directly using wired links. For this reason, we aim to interconnect all ToR switches according to a dedicated wireless network topology. In this chapter, we adopt the widely used Fat-Tree as an example of wired DCN and augment it with a wireless Torus topology. In this way, we achieve a hybrid VLCcube, which can seamlessly integrate both wired and wireless DCNs.

As depicted in Fig. 6.3, all racks in wired Fat-Tree DCN are further interconnected via VLC links to form a two-Dimensional wireless Torus, with m racks in each row and n racks in each column. On the top of each rack, four VLC transceivers are deployed towards four orthogonal directions, such that the interference can be restricted at the lowest level. Note that the wired part of the VLCcube is a Fat-Tree topology, and we just connect the ToR switches via VLC links. Let k denote the

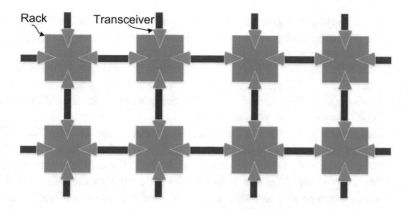

Fig. 6.3 The inter-rack wireless network of VLCcube. All racks in Fat-Tree are interconnected as a wireless Torus via VLC links

number of ports of each switch, which is usually even. Thus, VLCcube accommodates k pods, each of which has $k/2$ ToR switches and $k/2$ aggregation switches. Thus, $k^2/2$ ToR switches are involved in the wireless part of VLCcube.

In fact, we can design the hybrid VLCcube topologies in two ways. A straightforward way is to augment the wired network topologies with a wireless 2D Torus directly. By contrast, a more advisable method can further promote the topology by jointly consider the wireless 2D Torus and the wired Fat-Tree.

1. Independent topology design of wireless Torus

We notice that the wireless 2D Torus can be attached to the existing wired Fat-Tree directly. Without loss of generality, we assume that, in a k-pod Fat-Tree, there are k rows and $k/2$ columns, i.e., $m = k, n = k/2$. In each dimension, every rack enables wireless links with its neighboring racks. As a result, the diameter of the 2D Torus is $0.75k$, which is proportional to the number of k. Moreover, in the resulted wireless 2D Torus, given a rack in the i-th pod, each rack has four neighbors. Among these neighbors, two are in the same i-th pod, and the other two are in $i - 1$-th pod and $i + 1$-th pod. That is, in the pod level, the i-th pod just connects with $i - 1$-th pod and $i + 1$-th pod directly. This construction method of wireless Torus can increase the connectivity of racks and provide more alternative paths for servers. However, if we construct wireless Torus based on the deployment of racks, the gain of wireless Torus can be further improved.

2. Joint topology design of wireless Torus and wired Fat-Tree

To fully utilize the benefits of VLC wireless links, we optimize the topology of VLCcube by integrating the 2D Torus with Fat-Tree seamlessly. To reach this goal, two important issues must be well tackled, including the settings of m and n, and the placement of racks in VLCcube.

First, all racks in each dimension of Torus should be interconnected as a loop. Thus the network diameter of Torus is $(m + n)/2$. The VLCcube needs to set reasonable m and n to minimize the network diameter. Meanwhile, the number of long-range links in Torus is $m + n$. Considering that the data rate of long-range VLC links is less than that in short-range links, minimizing $m + n$ is also helpful for the improvement of network performance.

Under the structural rule of Torus, the placement of racks can be optimized. In Fat-Tree, the path length between any pair of racks is either 2 hops or 4 hops. To minimize the network diameter of VLCcube, the wireless links should be used to shorten these 4-hop paths as many as possible.

Parameter setting. Note that all nodes in each dimension of a 2D Torus form a loop structure; hence, the network diameter by $(m + n)/2$. For this reason, VLCcube aims to minimize the network diameter of the used Torus structure by inferring reasonable configurations of m and n. Additionally, the total number of remote VLC links in VLCcube is $m + n$, and such a few long links are more difficult to establish compared to those short VLC links. This issue further motivates VLCcube to minimize $m + n$ for eliminating unnecessary remote VLC links. If a 2D Torus is designed to accommodate $k^2/2$ racks, the parameters m and n should be bounded by the inequation $m \times (n - 1) < k^2/2 \leq m \times n$.

Theorem 6.1 *In VLCcube, the optimal setting of m is calculated as $\lceil \sqrt{k^2/2} \rceil$. The value of n depends on $k^2/2$. If $(m - 1)^2 < k^2/2 \leq m \times (m - 1)$, n is $m - 1$; in contrast, when $m \times (m - 1) < k^2/2 \leq m^2$, n is set the same as m, i.e., $\lceil \sqrt{k^2/2} \rceil$.*

Proof The best settings of m and n should minimize the value of $m + n$. Note that we have $m + n \geq 2 \times \sqrt{m \times n} \geq 2 \times \sqrt{k^2/2}$. Thus, $m + n$ reaches its minimum value only when $m = n$. Considering the inequation $m \times (n - 1) < k^2/2 \leq m \times n$, we derive the relationship between m, n and k. □

Placement of racks. As for the placement problem, we note that, during the design stage, the placement of racks and cables can be jointly optimized with the respect of wireless Torus topology. Indeed, the path length between any pair of ToRs is either two or four hops in Fat-Tree. Hence, VLCcube aims to shorten those four hops of wired communication as just one hop wireless communication by reconsidering the location of racks. Given m and n, we further concern about the best placement of racks for supporting the inter-rack wireless Torus network. Note that, in Fat-Tree, if two racks fall into the same pod, the path length between them is 2; otherwise, four hops are required. VLCcube targets shortening the four hops of the wired path as one hop wireless path. That is, all VLC links are utilized to connect those racks across pods, rather than those racks inside the same pod.

To ease the presentation of the placement strategy, we first introduce the identifier for each rack, which consists of two parts. The prefix, ranking from 0 to k, denotes which pod this rack belongs to. The suffix, ranking from 0 to $k/2$, identifies the rack in each pod. For example, the identifier 51 refers to the second rack in the sixth pod.

We further define the pod level logic graph, which regards a pod in VLCcube as a node. If there exist one or multiple VLC links between a pair of pods, an edge is added between them in the logic graph. Figure 6.4 depicts an example of the wireless part of VLCcube, with $k = 6$, $m = 5$ and $n = 4$. Accordingly, the pod level logic graph is derived. Typically, we measure the connectivity of the pod-level logic graph by

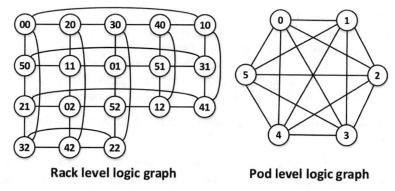

Rack level logic graph   **Pod level logic graph**

Fig. 6.4 The logic graph of VLCcube in the rack level and pod level

counting the number of links in the graph. In Fig. 6.4, 15 links interconnect six pods, thus the connectivity of the pod level graph is 15. Given the value of k for VLCcube, the connectivity of its pod level logic graph is no more than $k \times (k-1)/2$.

With the above definitions and given k, m and n, we design three steps to construct the 2D wireless Torus, which may be an incomplete one, as shown in Fig. 6.4.

- Step 1, allocating the prefixes. For each prefix $x \in [0, k]$, we randomly allocate it to $k/2$ ToR switches in the ToR level logic graph since each pod contains at most $k/2$ ToR switches. The only constraint is that any rack cannot hold the same prefix as its four neighbors. If conflicts occur, repeat this step until all prefixes have been mapped into the graph.
- Step 2, calculating the suffixes. In the ToR level logic graph, a suffix is introduced to differentiate those racks in the same pod. Note that the suffix of each rack ranges from 0 to $(k-1)/2$.
- Step 3, improving the connectivity of the pod level logic graph. We repeat the above two steps multiple rounds and then pick the solution that leads to the highest connectivity of the pod level logic graph.

However, as stated before, the Fat-Tree is actually installed as a w.l.o.g, $k^2/2$ array. Obviously, we need to transform the existing $k^2/2$ array to be a $m \times n$ one. Typically, two steps are needed. First of all, $(k-m) \times k/2$ racks must be moved such that the racks are placed as a $m \times n$ Torus physically. Then we deploy the placement strategy of racks logically by rewiring the cables between the aggregation switches and the racks. Undoubtedly, these adjustments suffer from dedicated time consumption and labor cost. But we believe these once-and-for-all augments are worthwhile to improve the network performance.

We further prove that our generation method can result in a correct VLCcube structure.

Theorem 6.2 *When $k \geq 4$, the above generation method can successfully generate a VLCcube such that each pod appears $k/2$ times in the ToR level logic graph.*

Proof In step 1, we allocate k pods randomly under the constraint that each link connects different pods. If each pod is associated with one color, the proof of Theorem 6.2 is equivalent to prove that k colors can color the graph successfully. In fact, the ToR level graph of VLCcube is a 4-regular graph whose chromatic number is 4, which means 4 kinds of colors are enough to color the graph. That is, when $k \geq 4$, we can always find out a legal placement strategy. Thus, Theorem 6.2 is proved. $\square$

Theorem 6.3 *The pod level logic graph resulting from the above three steps is connected.*

Proof Note that the ToR level logic graph is an incomplete 2D Torus, which is a connected graph. That is, a rack identified as $x_i y_i$ can find a path to its destination rack $x_j y_j$. If we map this path to the pod level logic graph, it is just the path from pod x_i to pod x_j. Thus, Theorem 6.3 is proved. $\square$

Theorems 6.2 and 6.3 ensure the rationality of the generation steps. Step 3 further ensures the connectivity of the pod level logic graph by selecting the best one after executing the first two steps multiple rounds. The behind insight is that by conducting the processes more rounds, we are more likely to achieve a better solution. We will evaluate the performance of such a generation method in further experiments.

From the view of topology design, VLCcube integrates the topological characteristics of both Fat-Tree and 2D Torus, e.g., scalability, constant degree, multi-path, and fault-tolerance. Moreover, VLCcube is easy-deployable and plug-and-play since only four transceivers are needed to deploy for each rack, and no further control operations are required during the usage process after the deployment. More importantly, VLCcube achieves the inter-rack wireless network without any modification to the hosting environment of a Fat-Tree data center.

6.3   Routing and Congestion Aware Flow Scheduling in VLCcube

For any pair of ToR switches, wired paths, wireless paths and hybrid paths coexist in VLCcube. The routing algorithms for wired paths and wireless paths can be found in Literatures [1, 10]. We focus on designing the hybrid routing between racks. To minimize the network congestion, we define a congestion-aware flow scheduling model and design scheduling algorithms for the batched and online traffic patterns.

6.3.1   Hybrid Routing Scheme in VLCcube

Given any pair of racks in VLCcube, the hybrid path Path$h$ consists of both wired links and wireless links. Therefore, the design of a hybrid routing algorithm must take into consideration the topological properties of both Fat-Tree and Torus. According to the characteristics of VLCcube, we design a top-down hybrid routing algorithm. Assuming the source and destination racks are $x_i y_i$ and $x_j y_j$ respectively. We first find the pod-level path from Pod x_i to Pod x_j. Then this path is completed at the rack level. In the practical process, VLC links are selected first, and then wired links are added to the hybrid path.

Firstly, the path from source Pod to destination Pod in the Pod level logic graph is calculated and is denoted as Path_{hp}. This step is relatively simple because there are only k nodes in the Pod level logic graph.

Secondly, each wireless link in Path_{hp} is obtained to construct the rack-level path Path_{ht}. Because there may be multiple optional wireless links between a pair of Pods, choosing different wireless links will lead to different lengths of the final routing path. Therefore, each hop in Path_{hp} should choose the wireless link that can make the routing path as short as possible. For example, in Fig. 6.4, select racks 11 and 41 as source rack and destination rack, respectively. In the Pod level logic graph, Pod 1 and Pod 4 are neighbors. In the rack level logic graph, there are 3 optional wireless links directly interconnect Pod 1 and Pod 4, namely, $12 \leftrightarrow 41$, $10 \leftrightarrow 41$ and $10 \leftrightarrow 40$. If link $10 \leftrightarrow 40$ is selected, racks 11 and 40 should firstly forward data to racks 10 and 41, respectively. Thus the length of the resulted path is 5 hops since both Pod 1 and Pod 4 need an aggregation switch to forward the inter-Pod data. However, if link $12 \leftrightarrow 41$ or $10 \leftrightarrow 41$ is selected, just 1 aggregation switch is needed to connect rack 11 to rack 12, or rack 11 to rack 10. In this case, the length of the resulted path is 4 hops.

At last, we add wired links to the generated Path_{ht}. Actually, this step is equal to add necessary aggregation switches to the path. In each pod, all ToRs and aggregation switches form a complete bipartite graph, thus we can pick an aggregation switch randomly.

Through the above three steps, we can construct the shortest hybrid path between any pair of racks. Note that the time complexity of step 1 is $O(k^2)$, and the time complexities of steps 2 and 3 are constants. Therefore the time complexity of this algorithm is $O(k^2)$, where k is the number of ports of the ToR switch. Since k is usually less than 100, $O(k^2)$ is an acceptable time complexity.

6.3.2   Problem Formulation of Flow Scheduling

We introduce the VLC links to augment the existing DCNs in VLCcube. The insight is to organize ToR switches as an incomplete wireless Torus via VLC links. To efficiently utilize both wired and wireless links and minimize the network's delay, we present a flow scheduling model to optimize the link congestion rate under both batched and online traffic patterns. Consider that there are four available transceivers on each rack. Thus, any rack can communicate with its four neighboring racks simultaneously. We first introduce related definitions and symbols as follows.

Let $G = (V, E)$ denote a data center network, where V and E are the node set and link set, respectively. Additionally, $F = \{f_1, f_2, \ldots, f_\delta\}$ denotes δ flows injected into

G. For each flow $f_i = (s_i, d_i, b_i)$, s_i, d_i, and b_i denote the source node, destination node and traffic demand, respectively. Typically, ϕ records a scheduling strategy, which is responsible to derive the routing path for each flow in F.

Definition 6.1 Given F and ϕ, we define the congestion rate of an arbitrary link e as:

$$C_F^\phi(e) = t(e)/c(e), \tag{6.1}$$

where $t(e)$ denotes the amount of traffic passing through link e, and $c(e)$ records the capacity of link e. Note that any $C_F^\phi(e)$ falls into a constant interval $[0, 1]$. Specifically, if none of flows passes through link e, its congestion rate is 0. The congestion rate is 1 when link e is fully used.

Definition 6.2 We define the congestion rate of a path P as

$$C_F^\phi(P) = \max C_F^\phi(e), \ \ where \ e \in P. \tag{6.2}$$

Accordingly, based on $C_F^\phi(P)$, we can locate the bottleneck in a given path and decide whether a path is capable of serving a given flow.

6.3.3 Scheduling the Batched Flows

Definition 6.3 (SBF: scheduling batched flows) Given $G(V, E)$ and a set of flow transmissions F, the goal of batched flow scheduling is to find a reasonable flow scheduling strategy ϕ^* such that $Z = \max_{e \in E} C_F^{\phi^*}(e)$ is minimized.

We accordingly formulate the SBF problem as follows:

$$\text{Minimize } Z$$

$$\sum_{f:f \in out(s_i)} t_f = b_i + \sum_{f:f \in in(s_i)} t_f \quad \forall i \tag{6.3}$$

$$\sum_{f:f \in in(d_i)} t_f = b_i + \sum_{f:f \in out(d_i)} t_f \quad \forall i \tag{6.4}$$

$$\sum_{f:f \in in(x)} t_f = \sum_{f:f \in out(x)} t_f \quad \forall i, \forall x \notin \{s_i, d_i\} \tag{6.5}$$

$$b_{min}/c_{max} \leq Z \leq b_{max}/c_{min} \tag{6.6}$$

$$\sum_i w_e^i \times \left\lceil \frac{b_i}{c(e)} \right\rceil \leq Z \quad \forall i, \forall e \in E \tag{6.7}$$

$$w_e^i \in \{0, 1\} \quad \forall i, \ \forall e \tag{6.8}$$

In the above formulation, i is an integer in the range $[0, \delta]$. Let $out(v)$ and $in(v)$ denote the set of the outgoing and incoming flows at node v in VLCcube, respectively. t_f refers to the size of flow f. The first three equations ensure that each flow just transmits along one path. Equation 6.6 determines the upper and lower bounds of Z, where b_{max} and b_{min} denote the max value and min value of each flow, respectively. Equation 6.7 ensures that the congestion rate of each link is no more than Z. w_e^i represents whether link e is occupied by flow f_i. $w_e^i = 1$ if flow f_i transmits through link e, and $w_e^i = 0$ otherwise.

The SBF problem is an Integer Linear Programming (ILP) problem, which is a well-known NP-hard problem. It cannot be solved in polynomial time. Thus, we design a lightweight algorithm to derive a reasonable solution. For any $f_i \in F$, we find out the three kinds of routing paths in VLCcube and denote them as $\mathscr{P}(f_i)$. Actually, $\mathscr{P}(f_i)$ contains $k^2/4$ wired paths, one hybrid path and one wireless path. To derive the flow scheduling strategy for F, we design a greedy heuristic algorithm based on the concept of congestion coefficient.

Definition 6.4 Given a set of flows F, each flow $f_i \in F$ has a set of candidate routing paths $\mathscr{P}(f_i)$. The congestion coefficient of a link $e \in E$, denoted as l_e, is the total amount of candidate paths passing through it under all flows in F.

Definition 6.5 For any routing path $P \in \mathscr{P}(f_i)$ of any flow f_i in F. The congestion coefficient of P, denoted as l_P, can be calculated as $l_P = \sum l_e$, where $e \in P$.

Indeed, the congestion coefficient of link e or path P indicates the probability that multiple flows employ it, respectively. Hence, l_P is an index for our greedy algorithm to decide whether flow f_i should select path P. To be specific, we should select the path with the least congestion coefficient among all paths in $\mathscr{P}(f_i)$.

Based on the congestion coefficient, Algorithm 6.1 shows the insight of the greedy strategy. For each flow, we first calculate its $k^2/4 + 2$ candidate routing paths. Then, the congestion coefficient of each link in VLCcube is derived. For each flow f_i, we calculate the congestion coefficient for each candidate routing path and choose the path with the least congestion coefficient to serve that flow. The algorithm takes $O(\delta \times (k^2 + k + 4))$ time-consumption to derive the candidate routing paths for all flows in F, and additional $O(\delta \times (k^2/4 + 2))$ time-consumption to decide which routing path should be utilized by each flow. Hence, the total computation complexity can be calculated as $O(\delta \times k^2)$.

The congestion coefficient of a link e means there are up to l_e flows that may employ the link. The congestion coefficient of a path P demonstrates that at most l_P flows may pass through at least one link along the path. If no scheduling strategy is utilized, any path $P \in \mathscr{P}(f_i)$ has the equal probability to be chosen to transmit f_i. Based on this insight, Theorems 6.4–6.6 prove the correctness of employing the congestion coefficient of a path as an index for Algorithm 6.1.

Algorithm 6.1 SBF-solution (S_{batch})

Require: Input the model of SBF problem.

1: Initialize S_{batch} as empty;

2: For each $f_i \in F$, derive $P(f_i)$;

3: Calculate the congestion coefficient of each link in VLCcube;

4: **for** $i < \delta$ **do**

5:    Calculate the congestion coefficient of each path in $\mathscr{P}(f_i)$;

6:    Select the path with the least congestion coefficient;

7:    Add the chosen path into S_{batch};

8:    Label the links on the chosen path as used;

9: **return** The solution of SBF problem S_{batch};

Theorem 6.4 *In VLCcube, given a flow $f_i \in F$, e is an arbitrary link in the network, the probability that e is utilized by f_i is:*

$$p_e^{f_i} = \begin{cases} = 0, & f_i \notin F_e \\ = l_e^{f_i}/(k^2/4 + 2), & f_i \in F_e \end{cases} \tag{6.9}$$

where F_e records the set of flows that may employ the link e, $l_e^{f_i}$ denotes the congestion coefficient of the link e caused by f_i, since more than one candidate routing paths of f_i may cover that link e.

Proof Note that, if no scheduling strategy is utilized, any path $P \in \mathscr{P}(f_i)$ has the equal probability to be chosen to transmit f_i. For flow f_i, if number of $l_e^{f_i}$ paths in $\mathscr{P}(f_i)$ pass through link e, we have $p_e^{f_i} = l_e^{f_i}/(k^2/4 + 2)$. Otherwise, flow f_i never utilizes that link, and the probability is 0. Thus, Theorem 6.4 is proved.   □

Theorem 6.5 *In VLCcube, for any flow $f_i \in F$, η counts the number of flows that pass through a link e, then we have:*

$$p_e^F(\eta = 0) = \prod_{f_i \in F}(1 - p_e^{f_i}) \tag{6.10}$$

$$p_e^F(\eta = 1) = \sum_{f_i \in F}\left[p_e^{f_i} \times \prod_{f_j \in F - f_i}(1 - p_e^{f_j})\right] \tag{6.11}$$

$$p_e^F(\eta \geq 2) = 1 - p(\eta = 0) - p(\eta = 1) \tag{6.12}$$

Proof Given a flow set F, such flows are independent for whether employ a link e or not. Hence, $p(\eta = 0)$ and $p(\eta = 1)$ can be calculated easily. Thus, Theorem 6.5 is proved.   □

Theorem 6.6 *Consider a flow $f_i \in F$, let η counts the number of flows that pass through a path $P \in \mathscr{P}(f_i)$, and $E(P)$ denotes the set of links along the path P. For any P, we have:*

$$p_P^F(\eta = 0) = \prod_{e_i \in E(P)} p_{e_i}^F(\eta = 0) \tag{6.13}$$

$$p_P^F(\eta = 1) = \frac{4}{k^2 + 8} \prod_{e_i \in E(P)} p_e^{F-f_i}(\eta = 0) + \sum_{e_s \in E(P)} \left[p_{e_s}^{F-f_i}(\eta = 1) \times \right.$$

$$\left. \prod_{e_j \in E(P)-e_s} p_{e_j}^{F-f_i}(\eta = 0) \right] \tag{6.14}$$

$$p_P^F(\eta \geq 2) = 1 - p_P^F(\eta = 0) - p_P^F(\eta = 1) \tag{6.15}$$

Proof For a path $P \in \mathscr{P}(f_i)$, $\eta = 0$ means none of flows passes any link in path P. While, $\eta = 1$ is resulted from two situations, i.e., only f_i occupies the path P, or one link in path P has been utilized by another flow $f_j \in F - f_i$. Thus, $p_P^F(\eta = 0)$ and $p_P^F(\eta = 1)$ can be calculated. □

According to Theorems 6.4–6.6 calculate the probability that none or one flow passes link e and path P. Note that, if $\eta \geq 2$, link e or path P may result in congestion. This will happen when the completion time of the former flow blocks the transmission of the latter flow. Theorems 6.5 and 6.6 demonstrate that larger l_e leads to more opportunities that more than 2 flows go through link e or path P, and may cause congestion. Thus the probability that a path P is blocked is proportional to its congestion coefficient l_P. In this way, the correctness of employing the congestion coefficient of a path as an index for Algorithm 6.1 is certified. Since our greedy algorithm selects the paths with the least congestion coefficient, the congestion rate in VLCcube will be decreased significantly.

6.3.4   Online Scheduling of Flows

As discussed in [19], flows are not always batched in data centers. In fact, flows are usually uncertain and dynamic. Typically, ϕ^0 depicts an existing flow scheduling strategy, F_N denotes the new arriving flows, and F_O contains the flows that call for retransmission. Accordingly, we update the set of flows as $F_1 = F_N + F_O$. With F_1 as input, we define the online flow scheduling problem as follows:

Definition 6.6 (*SOF: scheduling online flows*) The SOF problem is to deduce a new scheduling strategy ϕ_1 such that the increased link congestion rate is minimized. Let $\Delta Z = Z_1 - Z_0$, where $Z_1 = \max_{e \in E} C_{F_1}^{\phi_1}(e)$ and $Z_0 = \max_{e \in E} C_{F-F_O}^{\phi^0}(e)$, the goal of SOF is to minimize ΔZ.

The SOF problem will be triggered when new flows appear, or some existing flows are required to be retransmitted. Note that the SOF problem still subjects to an ILP model, which is similar to the SBF problem. We omit the detailed presentation of the SOF model due to the page limitation.

The SOF problem targets at minimizing Z_1. Thus, it seems that the same strategy, depicted in Algorithm 6.1, can be utilized to solve the SOF problem. Algorithm 6.1, however, will be employed frequently due to the dynamic flows, and hence causes unnecessary computation cost. Instead, we only take flows in F_1 into consideration and propose a greedy flow scheduling strategy for the SOF problem. For each flow in F_1, the insight of our greedy strategy is to employ the path that causes the least link congestion rate.

Algorithm 6.2 SOF-solution (S_{online})

Require: Input the model of SOF problem.
1: Initialize S_{online} as empty;
2: Calculate the updated routing requests F_1;
3: Update the state of network links and devices;
4: **for** $i < \delta_1$ **do**
5:      Search the three kinds of paths from s_i to d_i;
6:      Calculate the congestion rate of each path;
7:      Select the path with the least congestion rate, i.e., $path_i$;
8:      Add $path_i$ into S_{online};
9: **return** The solution of SOF problem S_{online};

As depicted in Algorithm 6.2, the greedy strategy discovers those flows that call for path assignment. Typically, it distinguishes the finished flows, the new flows and the failed flows. The algorithm has to know which available links and devices the updated flows can employ. For each flow in F_1, Algorithm 6.2 searches all of its wireless path, hybrid path and wired paths. After deriving the possible candidate routing paths for a flow f_i, we calculate the congestion rate of each path according to the values of b_i and c_i. We then pick the path with the least congestion rate as the final routing path for f_i and add it into S_{online}. When each flow in F_1 has been assigned a reasonable path, Algorithm 6.2 returns the result. The algorithm will be executed δ_1 rounds since there are δ_1 flows in F_1, while the time consumption is $O(k + k^2)$ in each round, due to deriving the candidate paths. Thus, the computation complexity of Algorithm 6.2 is $O(\delta_1 \times k^2)$.

Theorem 6.7 *Algorithm 6.2 outperforms the traditional ECMP flow scheduling strategy for the online traffic pattern.*

Proof For any flow $f_i \in F_1$, if ECMP is employed, the expectation of congestion rate for f_i is

$$\frac{4}{k^2 + 8} \sum_{P_j \in \mathscr{P}(f_i)} C_{F_1}^{\phi^*}(P_j) \tag{6.16}$$

By contrast, the congestion rate of Algorithm 6.2 for f_i is

$$min\{C_{F_1}^{\phi^*}(P_j)\} \ s.t. \ P_j \in \mathscr{P}(f_i) \tag{6.17}$$

Undoubtedly, for an arbitrary flow f_i, the congestion rate under Algorithm 6.2 is no more than that under ECMP. Thus, Theorem 6.7 is proved.   □

6.4   Performance Evaluation

In this section, we first introduce the settings and methodologies. Then, we compare VLCcube with Fat-Tree, in terms of the topological properties and network performance. Finally, the proposed congestion-aware scheduling methods are evaluated against the widely used ECMP.

6.4.1   Settings of Evaluations

We realize the proposed VLCcube and Fat-Tree with Network Simulator (NS3). Given the number of k, we generate Fat-Tree according to the rules introduced in Literature [1]. As for VLCcube, we generate it with the steps proposed above. The bandwidth of each wired link and short-range VLC link is set as 10 Gbps, while the bandwidth of the long-range VLC link is set as 100 Mbps. With the above basic setting, we first compare their topological characteristics and then evaluate the complexity of routing algorithms for wired paths, wireless paths and hybrid paths. Moreover, we compare their network performance.

In our evaluations, we consider three traffic patterns: (1) Trace flows: the flows are generated by a real data-set from Yahoo!'s data centers [20]; (2) Stride-i flows: a server with id x sends the flow to the destination with id $(x + i) \ mod \ N$, where N is the total number of servers; and (3) Random flows: the source and destination of each flow are chosen randomly. The network throughput and packet loss rate are used to measure the performance of DCNs under diverse traffic patterns. Note that the arrival time of dynamic flows follows a Poisson distribution in the case of online flow scheduling.

We evaluate the network performance of VLCcube and Fat-Tree, which both utilize the prior ECMP flow scheduling method. Under each of the three traffic patterns, we vary the network scale by adjusting the value of k from 6 to 60, and observe the changing trends of the network throughput and packet loss rate. To reveal the impact of flow size on the network performance, the average size of each flow ranges from 5 to 300 Mb. Note that the size of each flow under the trace-based traffic pattern is always set according to the trace. In the experiments where k varies, the network throughput is normalized with the throughput measured when $k = 60$. In the experiments where the flow size varies, the network throughput is normalized with the throughput measured when the flow size equals 300 Mb.

6.4.2   Topological Properties of VLCcube

Two topological properties, the average path length (APL) and the total network
bandwidth are measured for VLCcube and Fat-Tree. As shown in Fig. 6.5a, b, VLC-
cube has shorter APL and offers higher network bandwidth than Fat-Tree, due to
those VLC wireless links. Note that the impact of VLC wireless links on the APL
exhibits an obvious marginal effect. That is, VLC wireless links for small-scale networks would significantly decrease the APL. In fact, given k, there are k^2 VLC links
in VLCcube, and the number of both wired and wireless links is $k^3/2 + k^2$. With the
increase of k, VLC links contribute less portion to the total number of links. Hence,
the impact of wireless links becomes weak.

   We conduct the generation progress multiple rounds to deduce the placement strategy for VLCcube. Then the connectivity of the pod level logic graph is normalized
and measured as the number of links in a complete graph. In Fig. 6.5c, $VLCcube_1$,
$VLCcube_2$ and $VLCcube_{10}$ denote the measured results under the picked placement
strategy with the highest pod level connectivity after one round, two rounds, and ten
rounds generations. It is clear that the connectivity reduces along with the increase of
k, irrespective of the rounds of generation. Additionally, $VLCcube_{10}$ always outper-

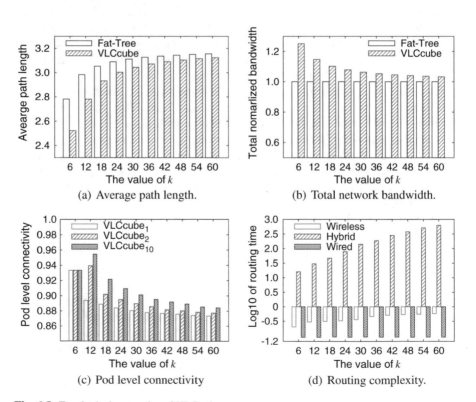

Fig. 6.5 Topological properties of VLCcube

forms the other two cases since the generation algorithm may derive a better solution from more candidates with high probability.

We further measure the consumed time due to calculating three kinds of routing paths in VLCcube. As depicted in Fig. 6.5d, the time-consumption of searching the hybrid path increases with k, and is always larger than that of the other two kinds of paths. The time consumption of searching the wireless paths also increases with k. Specifically, it grows from 0.2 to 0.575 ms. In addition, searching the wired paths consumes the least time, which stables at a very low level. Overall, the time complexity of the wired path routing algorithm is a constant, while the other two are proportional to k and k^2 respectively.

The above experimental results show that VLCcube provides better network performance, including higher network bandwidth and lower average path length.

6.4.3 Network Performance Under Trace Flows

Yahoo!'s trace records the basic information for each flow in its six distributed data centers, including the IP addresses of both source and destination servers, the flow size, the utilized interfaces, etc. We separate those inner data center flows from those flows across data centers by identifying the utilized interfaces [20]. We inject k^3 randomly chosen flows into the VLCcube and Fat-Tree networks to evaluate their performance.

Figure 6.6a, b plot the performance of VLCcube and Fat-Tree in terms of both throughput and packet loss rate when k varies from 6 to 60. It is clear that VLCcube dominates Fat-Tree by offering more throughput (8.5% more) and causing a much lower packet loss rate (39% less). The reason is that those VLC links provide more candidate paths for each flow.

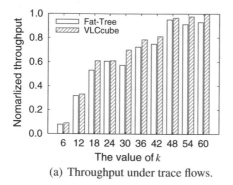
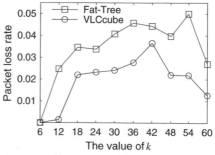

(a) Throughput under trace flows.　　(b) Loss rate under trace flows.

Fig. 6.6 Network performance under trace flows

6.4.4   Network Performance Under Stride-2k Flows

In this experiment, firstly, we set the average flow size as 150 Mb. We measure the throughput and packet loss rate under diverse network orders by increasing k from 6 to 60. Figure 6.7a, b report the evaluation results, when k^3 flows are injected in the networks. With the increase of k, both Fat-Tree and VLCcube are capable of accommodating more flows, thus their throughputs increase rapidly. But VLCcube achieves 15.14% more throughput than Fat-Tree on average, while the packet loss rate of VLCcube is much less than that of Fat-Tree.

To measure the impact of flow size, we vary the average size of those k^3 flows from 50 Mb to 300 Mb while $k = 36$. We can infer from Fig. 6.7c, d that VLCcube still considerably outperforms Fat-Tree. More precisely, VLCcube increases the network throughput up to 14.31% than Fat-tree even when the maximum flow size is 150 Mb. Reasonably, when the flow size grows, the packet loss rate increases dramatically.

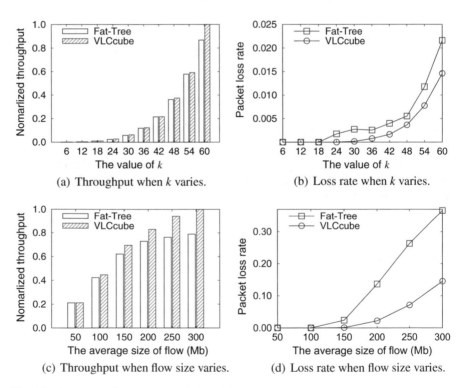

(a) Throughput when k varies.

(b) Loss rate when k varies.

(c) Throughput when flow size varies.

(d) Loss rate when flow size varies.

Fig. 6.7   Network performance under stride-2k flows

6.4.5 Network Performance Under Random Flows

In the random traffic pattern setting, the source and destination server of each flow are all selected randomly. We also introduce k^3 flows into the two data center networks.

First of all, we fix the average flow size as 150 Mb and vary the network scale by ranging k from 6 to 60. As shown in Fig. 6.8a, for both VLCcube and Fat-Tree, the network throughput increases dramatically. VLCcube still outperforms Fat-Tree with 10.44% more network throughput on average. As depicted in Fig. 6.8b, Fat-Tree always experiences a high packet drop rate when $k \geq 18$, while VLCcube incurs much less packet drop rate. To be specific, VLCcube and Fat-Tree drop 0.27 and 2.45% packets on average, respectively.

We further evaluate the impact of maximum flow size on the network performance when $k = 36$. As shown in Fig. 6.8c, d, VLCcube and Fat-Tree must transmit more packets once the flows are scheduled; hence, the throughput and loss rate increase reasonably as larger flows are injected into the networks. Compared with Fat-Tree, VLCcube has better network performance since it achieves higher network throughput and a lower loss rate.

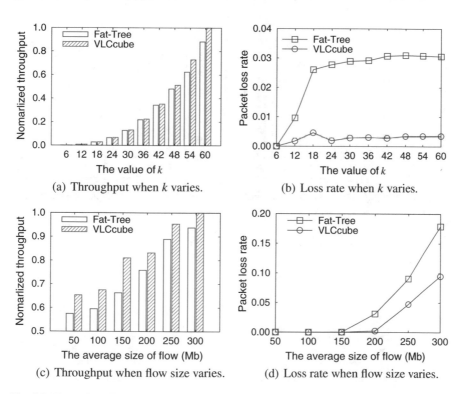

(a) Throughput when k varies.   (b) Loss rate when k varies.

(c) Throughput when flow size varies.   (d) Loss rate when flow size varies.

Fig. 6.8 Network performance under random flows

In summary, VLCcube achieves better network performance than Fat-Tree under the three kinds of traffic patterns when both of them employ the ECMP to schedule flows.

6.4.6   Impact of Congestion Aware Flow Scheduling

Although the above evaluations demonstrate the benefits of VLCcube over Fat-Tree, the topological benefits have not been fully exploited by employing the existing ECMP flow scheduling method. Thus, we compare our congestion-aware scheduling method with ECMP under different sizes of VLCcube.

We inject k^3 batched random flows into VLCcube, where k varies from 6 to 60. The network throughput is normalized as the ratio of the network throughput under the ECMP method to that under our scheduling method. As depicted in Fig. 6.9a, b, ECMP offers less network throughput and causes a worse packet loss rate. By contrast, our SBF method contributes $\times 1.54$ throughput and causes a much lower loss rate than ECMP. Note that the loss rate decreases dramatically when k increases

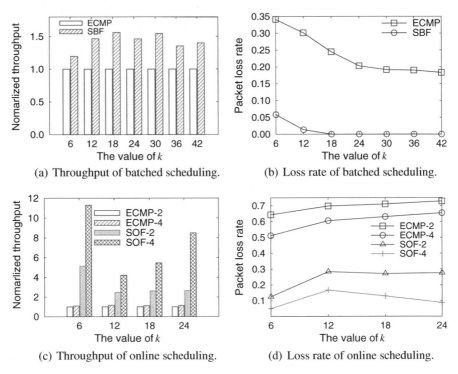

(a) Throughput of batched scheduling.   (b) Loss rate of batched scheduling.

(c) Throughput of online scheduling.   (d) Loss rate of online scheduling.

Fig. 6.9 The performance of congestion aware scheduling

from 6 to 18. The root cause of the low loss rate is that our SBF method offers more candidate paths and disperses the flows as widely as the VLCcube can.

Additionally, the VLCcube schedules online flow using our SOF method. In this case, we vary k from 6 to 24 and schedule k^3 random flows under each configuration of VLCcube. The arrival time of dynamic flows follows a Poisson distribution, whose parameter λ can be adjusted. Fig. 6.9c, d plot the evaluation results. Note that ECMP-x and SOF-x refer to the results under the ECMP and our SOF methods when $\lambda = x$. Our SOF method leads to $\times 2.22$ and $\times 5.56$ throughput than ECMP, while causes only $\times 0.340$ and $\times 0.178$ packet loss, when $\lambda = 2$ and $\lambda = 4$, respectively. Note that both ECMP and SOF methods in the case of $\lambda = 4$ outperform that in the case of $\lambda = 2$. The reason is, fewer flows will simultaneously arrive in a given time interval as the increase of λ; hence, such flows cause less packet loss.

Consequently, our SBF and SOF flow scheduling methods can improve the performance of VLCcube and realize less congestion rate than the widely used ECMP method.

6.5 Discussion

In this chapter, we augment the existing wired Fat-Tree DCN by introducing the inter-rack wireless network using VLC links. To fully understand the designing methodology of VLCcube, we discuss the following important issues.

The building methods of VLCcube. In this chapter, we point out that VLCcube can be constructed in two different methods. Indeed, the racks in an existing Fat-Tree data center have been placed as a fixed array, e.g., $k^2/2$ array. Hence, a simple method is to maintain the arrangement of racks and interconnect all racks as a wireless Torus. However, the resultant hybrid DCN brings limited improvement of network performance. To fully exploit the benefits of the introduced VLC links, we prefer to upgrade all racks from the $k^2/2$ array to a $m \times n$ array, which brings extra adjustment of racks and links. Consequently, the resultant VLCcube can replace more four-hop wired paths with one-hop wireless paths. Additionally, the connectivity among pods is enhanced by those VLC links. In summary, there is a trade-off between the extra cost and the promotion of performance for the second method.

The scalability of VLCcube. It is true that the Fat-Tree DCN lacks high scalability. Fat-Tree does not support incremental expansion, and its network order is decided by the value of k with the upper bound of $k^3/4$. When the network order is less than $k^3/4$, the resulted Fat-Tree is over-provisioned and leads to a lot of unnecessary device investment. On the other hand, if the demand on the network scale is larger than $k^3/4$, there can be some devices and links overload. To expand the network needs to increase the value of k. Fortunately, the wireless Torus can achieve the incremental expansion of network order since the wireless Torus and Fat-Tree are relatively independent. That is, the expansion of VLCcube can be realized by introducing racks to Torus directly. From this view, the wireless Torus can alleviate the expansion problem faced by Fat-Tree.

Rethinking of flow model and scheduling algorithms. In VLCcube, due to the existence of multiple paths, it is important to select a proper path for each flow since those paths result in diverse completion times. From the global view, we need to derive a suitable path for every flow, such that the congestion rate is minimized. After allocating the given path for each flow, more accurate and fine-grained control mechanisms can be realized by several existing proposals, e.g., Hedera, pFabric, L2DCT, etc. Such transport control mechanisms target optimizing the flow completion time, using dedicated rules, such as the shortest remaining processing time first, the deadline first, and the smallest flow first, etc.

The evaluation methodology. Our evaluations concentrate on measuring the impact of introducing VLC wireless links into the wired Fat-Tree networking structure for data centers. For fairness, we compare VLCcube with Fat-Tree when both of them employ the ECMP strategy. Then we evaluate the impact of the proposed scheduling algorithms for VLCcube. The comprehensive evaluations do verify the improvement of our VLCcube over Fat-Tree, in terms of both topological properties and network performance.

References

1. Al-Fares M, Loukissas A, Vahdat A. A scalable, commodity data center network architecture [J]. ACM SIGCOMM Computer Communication Review, 2008, 38(4): 63–74.
2. Greenberg A, Hamilton J R, Jain N, et al. VL2: a scalable and flexible data center network [J]. ACM SIGCOMM Computer Communication Review, 2009, 39(4): 51–62.
3. Hamedazimi N, Qazi Z, Gupta H, et al. FireFly: a reconfigurable wireless data center fabric using free-space optics [C]. In Proc. of ACM SIGCOMM, Chicago, USA, 2014: 319–330.
4. Zhou X, Zhang Z, Zhu Y, et al. Mirror mirror on the ceiling: Flexible wireless links for data centers [J]. ACM SIGCOMM Computer Communication Review, 2012, 42(4): 443–454.
5. Cui Y, Wang H, Cheng X, et al. Wireless data center networking [J]. IEEE Wireless Communications, 2011, 18(6): 46–53.
6. Shin J Y, Sirer E G, Weatherspoon H, et al. On the feasibility of completely wirelesss datacenters [J]. IEEE/ACM Transactions on Networking, 2013, 21(5): 1666–1679.
7. Guo C, Lu G, Li D, et al. BCube: a high performance, server-centric network architecture for modular data centers [J]. ACM SIGCOMM Computer Communication Review, 2009, 39(4): 63–74.
8. Guo C, Wu H, Tan K, et al. Dcell: a scalable and fault-tolerant network structure for data centers [J]. ACM SIGCOMM Computer Communication Review, 2008, 38(4): 75–86.
9. Singla A, Hong C Y, Popa L, et al. Jellyfish: Networking Data Centers Randomly [C]. In Proc. of 9th USENIX NSDI, Boston, USA, 2011: 17–17.
10. Shin J Y, Wong B, Sirer E G. Small-world datacenters [C]. In Proc. of 2nd ACM SOCC, Cascais, Portugal, 2011: 1–13.
11. Zhang W, Zhou X, Yang L, et al. 3D beamforming for wireless data centers [C]. In Proc. of 10th ACM HotNets, Cambridge, UK, 2011: 1–6.
12. Haas H. Light fidelity (Li-Fi): towards all-optical networking [C]. In Proc. of SPIE International Society for Optics and Photonics, 2013: 9007(5): 900702-900702-10.
13. Tsonev D, Chun H, Rajbhandari S, et al. A 3-Gb/s Single-LED OFDM-Based Wireless VLC Link Using a Gallium Nitride [J]. IEEE Photonics Technology Letters, 2014, 26(7): 637–640.
14. Chi Y, Hsieh D, Tsai C, et al. 450-nm GaN laser diode enables high-speed visible light communication with 9-Gbps QAM-OFDM [J]. Optics Express, 2015, 23(10): 13051–13059.

15. Ronja [EB/OL]. [2016-01-18]. http://ronja.twibright.com.
16. Singh S, Bharti R. 163m/10Gbps 4QAM-OFDM visible light communication [J]. IJETR, 2014, 2: 225–228.
17. PureLiFi [EB/OL]. [2016-01-18]. http://purelifi.com/lifi-products/li-1st/.
18. TracePro, http://www.lambdares.com/, 2015.
19. Han K, Hu Z, Luo J, et al. RUSH: RoUting and Scheduling for Hybrid Data Center Networks [C]. In Proc. of IEEE INFOCOM, Hongkong, China, 2015.
20. Chen Y, Jain S, Adhikari V K, et al. A first look at inter-data center traffic characteristics via yahoo! Datasets [J]. IEEE INFOCOM Proceedings, 2011, 2(3): 1620–1628.

Part III
Traffic Cooperation Management in Data Centers

Chapter 7
Collaborative Management of Correlated Incast Transfer

Abstract Data transfers, such as the common Shuffle and Incast communication patterns, contribute most of the network traffic in MapReduce like computing paradigms and thus have severe impacts on application performances in modern data centers. This motivates researchers to bring opportunities for performing the inter-flow data aggregation, during the transmission phase as early as possible rather than just at the receiver side. In this chapter, we first examine the gain and feasibility of the inter-flow data aggregation with novel data center network structures. To achieve such a gain, we model the Incast minimal tree problem. We propose two approximate Incast tree construction methods. We are thus able to generate an efficient Incast tree solely based on the labels of Incast members and the data center topology. We further present incremental methods to tackle the dynamic and fault-tolerant issues of the Incast tree. Based on a prototype implementation and large-scale simulations, we demonstrate that our approach can significantly decrease the amount of network traffic, save the data center resources, and reduce the delay of job processing. Our approach can be adapted to other data center topologies after minimal modifications.

7.1 Introduction

Large-scale data centers serve as infrastructures for not only online cloud applications, but also systems of massively distributed computing frameworks, such as MapReduce [1], Dryad [2], CIEL [3], Pregel [4], and Spark [5]. To date, such systems manage large number of data processing jobs each of which may utilize hundreds even thousands of servers in a data center. These systems typically follow a data flow computation paradigm, where massive data are transferred across successive processing stages. Such data transfers contribute 80% of the network traffic [5]; hence, they impose severe impacts on the application performance and the data center utilization. Hadoop traces from Facebook show that, on average, transferring data between successive stages accounts for one-third running time of a job on average in [6]. Prior work [6] aims at improving the data center utilization by carefully scheduling network resources for such data transfers. However, there has been relatively little work on directly lowering down the network traffic resulting from such data transfers.

© Springer Nature Singapore Pte Ltd. 2022
D. Guo, *Data Center Networking*,
https://doi.org/10.1007/978-981-16-9368-7_7

Many state-of-the-practice systems already apply aggregate functions at the receiver side to reduce the output data size. For example, during the Shuffle phase of a MapReduce job, each reducer is assigned a unique partition of the key range produced by the map stage. The reducer then pulls the content of its partition from every mapper's output and then performs the reduce function. For most reduce functions, e.g., MIN, MAX, SUM, COUNT, TOP-K, and KNN (K-nearest neighbors), they are associative and commutative. In such settings, as prior work [7] has shown, the reduction in size between the output data of all mappers and that of all reducers after aggregation is 81.7% for Facebook jobs. Such observations motivate this study where we consider applying the data aggregation to data transfer flows during the transmission process as early as possible rather than just at the receiver edge. If the inter-flow aggregation can be achieved, we will be able to significantly reduce the traffic and data center resource usage. The job can be speeded as well since the final input data to each reducer will be considerably decreased.

Currently, an Incast transfer is implemented by a set of unicast transmissions that do not account for collective behaviors of flows. Such a method brings less opportunity for efficient data aggregation. For an Incast transfer, the gain of inter-flow data aggregation can be achieved only if the involved flows can be cached and processed at some rendezvous nodes on their routing paths. In a switch-centric data center [7–12] using traditional switches, the computing ability and buffer space available on a commodity switch is very limited. Therefore, it is impractical for such kinds of structures to support the inter-flow data aggregation. Fortunately, many server-centric network structures [13–17] have been proposed for future data centers. They put the interconnection intelligence on servers and use switches only as crossbars; hence, they possess the capability of in-network cache and packet processing. Additionally, they provide multiple disjoint routing paths for any pair of servers and bring vast opportunities to managing the Incast transfer so as to form rendezvous nodes for inter-flow aggregation.

In this chapter, we examine the possible gain and feasibility of the inter-flow data aggregation for an efficient Incast transfer in server-centric data centers [5]. To maximize the achievable efficiency, the Incast transfer aggregation is formalized as the problem of constructing Incast minimal tree for directly lowering down the network traffic. We propose two approximate methods, RS (Routing Symbol)-based and ARS (Advanced Routing Symbol)-based methods, to build an efficient Incast tree. On this basis, we consider the dynamic and fault-tolerant issues in building the Incast tree and propose incremental solutions.

We evaluate the proposed approach with a prototype implementation and large-scale simulations. The results demonstrate that our approach can significantly reduce the network traffic, save data center resources, and speed up the computation of a job. More precisely, our ARS-based approach saves traffic by 38% on average for a small-scale Incast transfer with 320 senders in data centers BCube$(6, k)(2 \leq k \leq 8)$. It saves traffic by 58% on average for Incast transfers with 100–4000 senders in a data center BCube$(8, 5)$ with 262,144 servers. Even more network traffic can be saved if the members of an Incast transfer in the data centers are selected in a structured way instead of random.

7.2 In-Network Aggregation of an Incast Transfer

We use the MapReduce application as an example to analyze the common Shuffle transfer in data centers. A MapReduce job consists of two successive stages. In the "map" stage, each of the map tasks applies a map function to each input record and generates a list of key-value pairs. In the "reduce" stage, every reduce task applies a user-defined reduce function, usually an aggregative function, to each input record. Each reducer is assigned a partition of the key range produced by the map stage and pulls the content of its partition from every mapper's output during the Shuffle phase. The data flows from two mappers to the identical reducer are highly correlated since the key range and its partitions are the same for all mappers. Recently, there has been a growing interest in supporting data pipelining during the Shuffle phase. To this end, each mapper is modified to push data to reducers [18–20].

No matter a MapReduce job utilizes the pull or push mechanism, a Shuffle transfer is established between the map and reduce stages. Other distributed computing frameworks, like Dryad, CIEL, Pregel, and Spark, possess similar constructs. In general, a Shuffle consists of m senders and n receivers where a data flow is established for every pair of a sender and a receiver. An Incast consists of m senders and one of n receivers, where a data flow is established from every sender to the same receiver.

The Shuffle transfer can be decomposed as a set of independent Incast transfers each with the same senders to each individual receiver, thus, this chapter focuses on the data aggregation in Incast transfer. We will examine the gain and feasibility of the inter-flow data aggregation for an efficient Incast transfer. We then discuss the problem of constructing an Incast minimal tree as well as how to perform the inter-flow data aggregation on the tree.

7.2.1 Feasibility of the Inter-flow Data Aggregation

We observe that the data flows in an Incast transfer, are highly correlated during the Shuffle phase. The major reason is that the key range and its partitions are identical for all mappers. The data, a list of key-value pairs, among such flows thus share the identical key partition allocated to the same receiver. Therefore, for a key-value pair in one flow of an Incast transfer there exists a key-value pair with the identical key in any other flow with high probability, as shown in Fig. 7.1.

The key insights of this chapter are two-fold: (1) the receiver typically applies an aggregation function, e.g., MIN, MAX, SUM, COUNT, TOP-K, and KNN, to the received data from all flows in such an Incast transfer. (2) although the combiner at each sender has aggregated its output before delivering to the receiver, there exist considerable opportunities for the data aggregation between different flows. Such insights motivate us to think whether we can apply the same aggregate function to flows in the Incast transfer during the transmission phase as early as possible

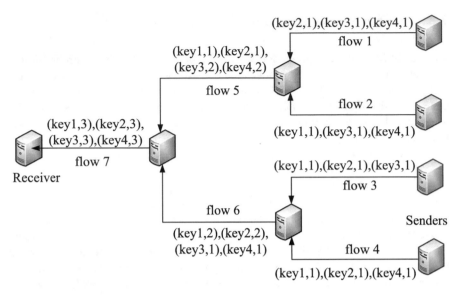

Fig. 7.1 An example of the inter-flow data aggregation for an Incast transfer

instead of waiting until they reach at the receiver side. If such an inter-flow data aggregation can be applied, it would not hurt the correctness of the computing result at the receiver.

For an Incast transfer, the gain of inter-flow data aggregation can be achieved only if its flows have opportunities to be cached and processed at some network devices in their physical paths. For those switch-centric structures, a traditional switch usually does not deliver programmable data plane although the software-defined networking technologies make the programmable control plane become possible. Therefore, pushing user-defined aggregation functions to traditional switches is prohibitive; hence, those switch-centric structures cannot naturally support the inter-flow data aggregation. Fortunately, new Cisco ASIC and Arista application switches have provided a programmable data plane, making it possible to perform inter-flow data aggregation in switch-centric structures.

For the server-centric structures of data centers, the commodity servers each with multi-ports act as not only end hosts, but also mini-switches. In practice, a server can connect a ServerSwitch card [21] that uses a gigabit, programmable switching chip for customized packet forwarding. A ServerSwitch card can leverage the server CPU for advanced in-network packet processing due to the high throughput and low latency between the switching chip and server CPU. As prior work [21, 22] have shown, a server with a ServerSwitch card can enable new data center networking services, e.g., in-network packet cache. If some flows in an Incast transfer intersect at a server, the early arrived packets can be cached on the server for the inter-flow data aggregation.

7.2.2 Minimal Incast Tree

For a data center, we model it as a graph $G = (V, E)$ with a vertex set V and an edge set E. A vertex of the graph corresponds to a switch or a datacenter server. An edge (u, v) denotes a link, through which u connects with v where $v, u \in V$.

In this chapter, our objective is to minimize the amount of the network traffic for completing an Incast transfer by applying the inter-flow data aggregation for the correlated flows. Given an Incast transfer with a receiver R and a set of senders $\{s_1, s_2, \cdots, s_m\}$ in a data center, we need to form an Incast tree from the graph $G = (V, E)$. The data flow from each sender thus can be delivered to the receiver r along the formed tree. There exist so many such trees for the Incast transfer in densely connected data center networks, e.g., BCube and FBFLY. A challenging issue is how to identify an Incast tree that results in less amount of network traffic after applying the inter-flow data aggregation.

For any Incast tree, a vertex can achieve the inter-flow data aggregation if it enables the in-network cache and receives at least two incoming flows. Such kinds of vertices are called aggregating vertices, and others are the non-aggregating ones. Note that the generated flow at a vertex acts as an incoming flow to it. At each aggregating vertex, multiple data flows consisting of key-value pairs can be aggregated as a new one. To ease the explanation, we first assume that the size of the outgoing flow of an aggregating vertex is the maximum size among its all incoming flows. We further evaluate a more general Incast transfer in the experiment section. At a non-aggregating vertex, the size of its outgoing flow is the cumulative size of its incoming flows since it cannot support the in-network cache and advanced packet processing.

Given an Incast tree, we define its cost metric as the amount of introduced network traffic for completing the Incast transfer. More precisely, the cost of an Incast tree is the total edge weight, i.e., the sum of the amount of the outgoing traffics of all vertices in the Incast tree except the receiver. Without loss of generality, the size of traffic resulting from each of the m senders is assumed to be 1 MB so as to normalize the cost of an Incast tree. In such a way, the weight of the outgoing link is one at an aggregating vertex and equals to the number of incoming flows at a non-aggregating vertex.

Definition 7.1 For an incast transfer, the minimal Incast tree problem is to find a connected subgraph, in $G = (V, E)$, that spans all Incast members with minimal cost for completing the Incast transfer.

The problem is then translated to how we discover an minimal Incast tree for an Incast transfer in a data center.

7.2.3 Inter-flow Data Aggregation on an Incast Tree

For any Incast transfer, an Incast manager is needed to ensure that all flows in this Incast is delivered along the generated Incast minimal tree, as described in the following steps:

First, the Incast manager makes all servers in the Incast tree be aware of the tree structure by broadcasting the resultant Incast tree to them. Each involved server in the Incast tree thus knows its parent server and its sub-tree (if any). A server can be an aggregating one if it has more than one child servers or it is a sender and has one child. For example, the servers v_0, v_1, and v_2 are aggregating servers, as shown in Fig. 7.2c. The job manager creates some data aggregation tasks, and places these tasks at such aggregating servers, then these servers will perform data aggregation, before sending the aggregated data downstream. Such a strategy will be much easier to be deployed than the network level approach.

Each sender of the Incast transfer delivers a data flow toward the receiver along the Incast tree if its output for the receiver has been generated and it is not an aggregating server. If a data flow meets an aggregating server, all packets will be cached. Upon the data from all its children and itself (if any) have been received, an aggregating server performs the inter-flow data aggregation as follows. It groups all key-value pairs in such flows according to their keys and applies the aggregate function used at the receiver to each group. Such a process finally replaces all original flows with a new one that is continuously forwarded along the Incast tree.

An aggregating server may perform the aggregation operation once a data flow arrives. Such a scheme amortizes the delay due to wait and simultaneously aggregate all incoming flows from itself and its children. At the root of the tree, i.e., the receiver of the Incast transfer, all of received data are aggregated using the reduce function. Figure 7.1 depicts an example of the inter-flow data aggregation based on an Incast tree. We can see that flows 1 and 2 are merged as a new flow 5, flows 3 and 4 are aggregated as a new flow 6. The flows 5 and 6 are forwarded to the receiver and aggregated as a flow 7.

7.3 Efficient Building Method of an Incast Tree

We start with two approximate Incast tree building methods, the RS-based and ARS-based approaches. We then optimize the Incast tree construction and present incremental methods to tackle the dynamic and the fault-tolerant issues.

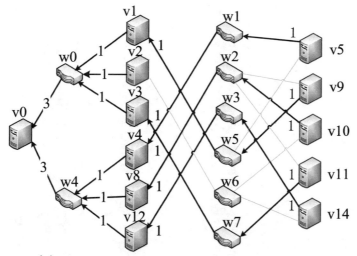

(a) A unicast-driven tree of cost 22 with 18 links.

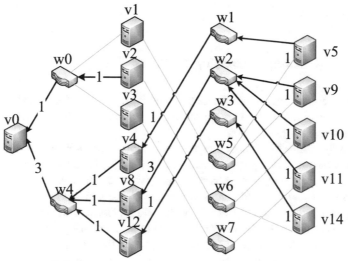

(b) An Incast tree of cost 18 with 14 links.

Fig. 7.2 Different incast trees for an incast transfer with the sender set $\{v_2, v_5, v_9, v_{10}, v_{11}, v_{14}\}$ and the receiver v_0 in a Bcube(4,1)

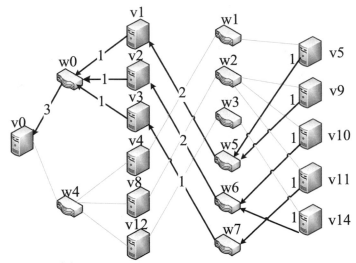

(c) An Incast tree of cost 16 with 12 links.

Fig. 7.2 (continued)

7.3.1 Online Construction of an Incast Tree

To meet the requirement of online tree building for Incast transfers with large number of senders, we aim to design an approximate method of exploiting the topological feature of data center networks.

In densely connected data center networks, there are multiple available unicast paths between any server pair. For example, BCube(n, k) has $k + 1$ equal-cost disjoint paths between any two servers if their labels differ in $k + 1$ dimensions. A straightforward solution to the Incast minimal tree problem is that each sender independently delivers its data to the receiver along a unicast path. The unicast-driven Incast tree is thus formed by the combination of such unicast paths. Such a method, however, has less chance of achieving the full gain of the inter-flow data aggregation.

Figure 7.2a plots an example of the unicast-driven Incast tree that we may construct in a BCube(4,1). The receiver is v_0 and the sender set is $\{v_2, v_5, v_9, v_{10}, v_{11}, v_{14}\}$. The resultant Incast tree is of cost 22 with 18 links and has no aggregating vertex. However, an efficient Incast tree for the same Incast transfer consists of only 12 links if we construct in the way shown in Fig. 7.2c. The total cost of the resultant Incast tree is 16 and there exist two aggregating vertices, v_1 and v_2.

The Incast manager controls the building of minimal Incast tree in a centralized manner. The JobTracker of a MapReduce job, which is aware of the senders and the receiver of each Incast transfer, notifies the Incast manager of an Incast transfer. Upon receiving an Incast transfer, the Incast manager calculates an efficient Incast tree using an approximate method, given the labels of all Incast members and the data center topology. The Incast manager is somewhat similar to the controller in software-defined networks [24, 25].

The topology of a BCube(n, k) can be abstracted as a $k + 1$ dimensional n-ary generalized hypercube [26] with the same set of servers. For a BCube(n, k) and the generalized hypercube, two servers $x_k x_{k-1} \ldots x_2 x_1 x_0$ and $y_k y_{k-1} \ldots y_2 y_1 y_0$ are called the mutual j dimension 1-hop neighbors if their labels differ in only the j dimension, where x_i *and* $y_i \in \{0, 1, \ldots, n - 1\}$ for $0 \leq i \leq k$. Therefore, a server has $n - 1$ 1-hop neighbors in each dimension. The difference is that such two servers directly connect with each other in the generalized hypercube while connect to a j level switch with a label $y_k \ldots y_{j+1} y_{j-1} \ldots y_1 y_0$ in a BCube(n, k). In such a case, a server and it's all neighbors in a dimension are connected indirectly via a common switch. Additionally, two servers are the mutual j-hop neighbors if their labels differ in number of j dimensions for $0 \leq j \leq k$.

Consider an Incast transfer consists of a receiver R and a set of senders $\{s_1, s_2, \ldots, s_m\}$ in a BCube(n, k). Assume that each receiver is labeled as $r_k r_{k-1} \ldots r_1 r_0$ and a sender is labeled as $s_k s_{k-1} \ldots s_1 s_0$, where $r_i \in \{0, 1, \ldots, n - 1\}$ and $s_i \in \{0, 1, \ldots, n - 1\}$ for $0 \leq i \leq k$. Without loss of generality, we assume that the maximum hamming distance between the receiver and every sender in the Incast transfer is $k + 1$ and discuss other cases later. Thus, the shortest paths from all senders to the receiver can be expanded as a directed multistage graph with $k + 2$ stages. The stage 0 only has the receiver. The servers at the stage j must be some of the j-hop neighbors of the receiver for $1 \leq j \leq k + 1$. Additionally, there must exist a set of switches as relays between two successive stages since servers are not directly connected in a BCube(n, k).

Note that, each of all senders should appear at the stage j if it is a j-hop neighbor of the receiver. For example, senders v_5, v_9, v_{10}, v_{11}, and v_{14} are at the stage 2 while another sender v_2 is at the stage 1, as shown in Fig. 7.2. Only such senders and the receiver in a BCube(n, k), however, cannot definitely form a connected subgraph in the level of a generalized hypercube. The problem is then translated as how to select some additional servers for each stage and switches between successive stages so as to constitute an Incast minimal tree.

Definition 7.2 For a server set A at the $j - 1$ stage and a server set B at the j stage in an Incast tree for $1 \leq j \leq k + 1$, we call A covers B if and only if for each server $u \in B$, there exists a directed path from itself to a server $v \in A$. We call A strictly covers B if any subset of A does not cover B.

Recall that a pair of neighboring servers across two successive stages connects to a common switch whose label and level can be derived from the labels of the two servers. For such reason, we only focus on the selection of additional servers for each stage. The insight behind Definition 7.2 is to ensure that a data flow from a server, in the j stage, toward the receiver can be definitely sent to a server in the $j - 1$ stage.

We identify other servers besides the senders at each stage from the stage $k + 1$ to the stage 1, recursively. A constraint must be satisfied to constitute an efficient Incast tree. The server set at any stage $j - 1$ strictly covers the server set at the successive stage j for $1 \leq j \leq k + 1$. Given the server set at the stage j, we leverage the topology features of a BCube(n, k) to infer the required servers at the stage $j - 1$ under such a constraint for $1 \le j \le k + 1$. Such server sets at the stage $j - 1$, however, may not be the only one in a BCube(n, k). The cause is that each server in the stage j has a mutual neighbor at the stage $j - 1$ in each of the j dimensions in which the labels of the current server and the receiver differ. Such a feature motivates us to define a routing sequence for an Incast transfer.

Definition 7.3 Let $e_1 e_2 \ldots e_j \ldots e_k e_{k+1}$ be a routing sequence, with $k + 1$ routing symbols. It is one of $(k + 1)!$ permutations of the set $\{0, 1, 2, \cdots, k\}$. The stage j of an Incast tree is associated with a routing symbol e_j for $1 \le j \le k + 1$. For each server $X = x_k x_{k-1} \cdots x_j \ldots x_1 x_0$ in the stage j and the receiver R,

1. If their labels differ in the dimension e_j, X selects one of its $n - 1$ mutual neighbors in the dimension e_j whose label is identical in the dimension e_j but different in $j - 1$ dimensions compared to the receiver's label.
2. Otherwise, their labels differ in at least one dimension from dimensions e_{k+1} to e_{j+1}. Let the rightmost such dimension be $\overline{e_j}$. Thus, the server X selects one neighbor in the dimension $\overline{e_j}$ whose label is identical in the dimension $\overline{e_j}$ compared to the receiver's label.

In such a way, the selected neighbor of the server X appears at the stage $j - 1$ in the Incast tree.

Given an Incast transfer and its associated routing sequence $e_1 e_2 \ldots e_j \ldots e_k e_{k+1}$. Its corresponding Incast tree under the above constraint along the processes as follows.

We start with any stage j in the $k + 2$ stages and assume that $j = k + 1$, without loss of generality. Upon the servers at the stage j are given, we partition all such servers into groups such that all servers in each group are mutual one-hop neighbors in the dimension e_i. In each group, the label of each server differs in j dimensions compared to the receiver's label, i.e., they are j-hop neighbors of the receiver. The next server along the path from each server in the group to the receiver should differ in $j - 1$ dimensions and appears in the $j - 1$ stage. Note that each server in the group has j selections about the next server towards the receiver. Here, we require that they share a common next server at the stage $j - 1$.

To this end, we have each server in a group selecting the next sever using the scheme mentioned in Definition 7.3. Thus, all servers in a group and their common next server have the same label except the e_j dimension. Actually, the e_j dimension of the common next server is just equal to the receiver's label at the e_j dimension. As shown in Fig. 7.2b, all servers at the stage 2 are partitioned into three groups, $\{5\}$, $\{9, 10, 11\}$, and $\{14\}$, using a routing symbol $e_2 = 0$. The next servers at the stage 1 of such groups are with labels 4, 8, and 12, respectively. In such a way, the data flows from the group toward the receiver can achieve the desired inter-flow data aggregation at the stage $j - 1$ if exists, as shown in Fig. 7.2b. Otherwise, there is less opportunity to do so at the stage $j - 1$, as shown in Fig. 7.2a.

After processing other groups in the stage j via the same method, the server set at the stage $j - 1$ that strictly covers the server set at the stage j are achieved. Note that some of senders may also appear at the $j - 1$ stage. So far, the set of all servers on the $j - 1$ stage are the union of the two parts. We can thus apply our approach to infer the required servers at the stage $j - 2$, and so on. Finally, all servers in all $k + 2$ stages and the directed paths between successive stages constitute an Incast tree for the given Incast transfer in a BCube(n, k). The resultant tree is called the routing sequence based Incast tree, abbreviated as the RS-based Incast tree.

Theorem 7.1 *For an Incast transfer with m senders in a BCube(n, k), the time complexity of the RS-based Incast tree building method is $O(m \times \log N)$, where $N = n^{k+1}$ refers the number of servers in a BCube(n, k).*

Proof Consider that our approach has to apply to at most k stages, from stage $k + 1$ to stage 2, in an Incast tree. The process to partition all servers at a stage j into groups according to e_j can be simplified as follows. For each existing server at the stage j, we extract its label except the e_j dimension. The resultant label denotes a group that involves such a server. We then add the server into such a group and infer the common next server of all possible servers in the group. The upcoming servers in the group need not calculate the common next server again. The computational overhead at the stage j is proportional to the number of servers at such a stage. Thus, the time complexity at each stage is $O(m)$ since the number of servers at each stage cannot exceed m. The time complexity of our approach is $O(m \times k)$ due to involve at most k stages. Thus, Theorem 7.1 proved.   □

7.3.2 Construction of the Minimal Incast Tree

A routing sequence $e_1 e_2 \cdots e_j \cdots e_k e_{k+1}$ can yield an RS-based Incast tree for any Incast transfer in a BCube(n, k). Note that there exist at most $(k + 1)!$ routing sequences for an Incast transfer, we can derive $(k + 1)!$ such Incast trees. They achieve different gains of the inter-flow data aggregation and have diverse tree costs. For example, we can derive an Incast tree under a routing sequence $e_1 e_2 = 10$ as shown in Fig. 7.2b, and another Incast tree under a routing sequence $e_1 e_2 = 01$, as shown in Fig. 7.2c.

A challenging issue is how to find an RS-based Incast tree with the minimal cost among $(k + 1)!$ candidates for an Incast transfer in a BCube(n, k). A brute force way is to generate all possible RS-based Incast trees by applying our RS-based method under each of the $(k + 1)!$ routing sequences. We then calculate the cost of each RS-based Incast tree and select the one with the minimal cost. Such a brute force way suffers high computation overhead with the time complexity of $O((k + 1)! \times k \times m)$.

We improve it by an efficient method as follows. We start with the stage $j = k + 1$. After defining a routing symbol $e_j (0 \leq j \leq k)$, all servers at the stage j can be partitioned into groups. From each group, we can identify an additional server for the stage $j - 1$ using our RS-based method. Thus, the number of partitioned groups is just equal to the number of appended servers at the next stage. The union of such appended servers and the potential senders at the stage $j - 1$ constitute the entire server set at such a stage. Recall that the outgoing flows from the stage j will be merged at the aggregating servers if exist, the number of outgoing data flows from the stage $j - 1$ is just equal to the size of the server set at such a stage. Inspired by such a fact, we apply our RS-based method to other k settings of e_j and accordingly achieve other k server sets that can appear at the stage $j - 1$. So far, we can simply select the smallest server set from all $k + 1$ ones. The related setting of e_j is marked as the best choice for the stage j among all $k + 1$ candidates. In such a way, the data flows from the stage j can achieve large gains of the inter-flow data aggregation at the next stage $j - 1$.

Upon the smallest server set at the $j - 1$ stage is given, we can achieve the smallest server set at the successive stage $j - 2$ and know the best choice of e_{j-1} at the stage $j - 1$, and so on. Finally, server sets at all $k + 2$ stages and the directed paths between successive stages constitute an Incast tree. The aforementioned routing sequence is simultaneously determined. Such a method is called the **ARS-based Incast tree building method**.

We use an example to reveal the benefit of our ARS-based approach. Consider the Incast transfer shown in Fig. 7.2. All senders at the stage 2 are partitioned into three groups, {5}, {9, 10, 11}, and {14}, using a routing symbol $e_2 = 0$. The resultant server set at the stage 1 is {2, 4, 8, 12}, as shown in Fig. 7.2b. In the case of $e_2 = 1$, all senders at the stage 2 are partitioned into three groups, {5, 9}, {10, 14}, and {11}, and the resultant server set at the stage 1 is {1, 3, 4}, as shown in Fig. 7.2c. Therefore, we definitely select the server set {1, 3, 4} for the stage 1 and $e_2 = 1$ for the stage 2. The selected Incast tree, as shown in Fig. 7.2c, has lower cost than the one, as shown in Fig. 7.2b.

Theorem 7.2 *Given an Incast transfer with m senders in a BCube(n, k), the time complexity of our ARS-based Incast tree building method is $O(m \times (\log N)^2)$, where $N = n^{k+1}$ refers the number of servers in a BCube(n, k).*

Proof Consider that our method has to apply to at most k stages, from stage $k + 1$ to stage 2, in an Incast tree. As shown in the proof of Theorem 7.1, the computation cost at the stage $k + 1$ for determining its server set is $O(m)$ given a routing symbol e_{k+1}. In the ARS-based building method, we conduct the same operations under all of $k + 1$ settings of e_{k+1}. This generates $k + 1$ server sets for the stage k at the cost of $O((k + 1) \times m)$ computation cost. Additionally, it incurs $O((k + 1) \times m)$ computation cost to identify the smallest server set from $k + 1$ candidates. In summary, the total computation cost of the ARS-based building method is $O((k + 1) \times m)$ at the stage $k + 1$.

At the stage k, the ARS-based approach can identify the smallest server set for the stage $k - 1$ from k candidates resulting from k settings of e_k since e_{k+1} has exclusively selected one from the set $\{1, 2, \ldots, k + 1\}$. The resultant computation cost is thus $O(k \times m)$. In summary, the total computation cost of the ARS-based building method is $O(\frac{k(k+3)}{2} \times m)$ since it involves at most k stages. Thus, Theorem 7.2 proved.   □

7.3.3   Dynamical Behaviors of Senders

When a sender s_{m+1} joins an existing Incast transfer with the senders $\{s_1, s_2, \ldots, s_m\}$ and a receiver r in a BCube(n, k), the Incast tree should be updated to embrace such a new sender. A brute force way is to generate a new Incast tree by invoking our ARS-based approach given the set of senders $\{s_1, s_2, \ldots, s_{m+1}\}$. Such a scheme, however, not only incurs $O(k^2 \times m)$ computation cost but also consumes network bandwidth for disseminating the new Incast tree to its all servers.

We prefer to update the Incast tree in an incremental way. First of all, the Incast manager keeps the routing sequence associated with the current Incast minimal tree. Upon it schedules a new sender s_{m+1} into the existing Incast transfer, it derives a unicast path from s_{m+1} to r by invoking our RS-based method at the cost of $O(k)$ computation overhead. The combination of the unicast path and prior Incast tree constitutes the final Incast tree for the new Incast transfer. To enable all servers in the final Incast tree be aware of such changes, the Incast manager only needs to propagate the unicast path structure to all servers along the path. In such a way, each server in the final Incast tree is aware of the subtree rooted at itself that is sufficient for performing the inter-flow data aggregation. Additionally, the path from each prior sender to the receiver r does not suffer any change.

When a sender s_j leaves the Incast transfer, the Incast tree should be updated to embrace such a change in an incremental way. First of all, the Incast manager extracts the unicast path from the sender s_j to the receiver r in the existing Incast tree. Note that extracting the unicast path for a leaving sender needs to decrease the weight of each related edge in the existing Incast tree by one and the edge with weight 0 will be removed from the tree. In such a way, we achieve the resultant Incast tree for the new Incast transfer. To enable all servers in the final Incast tree be aware of such changes, the Incast manager only needs to propagate the extracted path to all servers along the path. Such a process does not change the paths from other senders to the receiver.

In summary, our Incast tree building method can gracefully handle the dynamical behaviors of senders. In a MapReduce job, a master server schedules an idle server to replace a failed map task. Such a process consists of the leaving of a sender and the joining of a new sender.

7.3.4   Dynamical Behaviors of the Receiver

Similarly, the receiver r of an Incast transfer may be replaced with a new one, denoted as r_n. For example, a master server will schedule an idle server to replace a failed reduce task in a MapReduce job. In such a case, the Incast manager may generate a new Incast tree by using our ARS-based approach. Such a scheme, however, incurs $O(k^2 \times m)$ computation overhead and consumes network bandwidth for updating the Incast tree.

On the contrary, we prefer to update the Incast tree in an incremental way. First of all, the Incast manager keeps the routing sequence, $e_1 e_2 \ldots e_j \ldots e_k e_{k+1}$, associated with the Incast minimal tree for prior Incast transfer. Given prior receiver r and the new receiver r_n, we compare their labels along $k + 1$ dimensions, $e_1, e_2, \ldots, e_k, e_{k+1}$, one by one. If their labels differ only in the dimension e_j, the two Incast trees rooted at r and r_n and are equivalent from the stage $k + 1$ to the stage j but differ from the stage j to the stage 0. In other words, not only the server sets but also the directed paths across two stages are identical from the stage $k + 1$ to the stage j at the two trees.

Inspired by such a fact, the Incast manager can partially reuse prior Incast tree rooted at r and just recalculates the tree structure from stage j to stage 0 for efficiently generating the Incast tree rooted at r_n as follows.

1. Given the routing symbols, $e_1 e_2 \ldots e_j$, and the server set at the stage e_j, the tree structure from stage j to stage 0 can be calculated by using the RS-based method.
2. If r_n appears in prior Incast tree rooted at r, its subtree in prior Incast tree still exists in the Incast tree rooted at r_n but should start with stage 0 in the new one. The reason is the location change of r_n in the two trees.

We can conclude that the smaller the j such a method is more efficient. If r_n is just one neighbor of r in dimension e_1 ($j = 1$), the Incast manager only needs to accordingly adjust the directed paths across the stage 1 and the stage 0. Given an Incast tree rooted at r with a routing sequence $e_1 e_2 \ldots e_j \ldots e_k e_{k+1}$, our findings suggest that all of the $n - 1$ mutual neighbors of r in the dimension e_1 are the best substitutes if the receiver r needs to be replaced. If none of such substitutes are idle, the Incast manager will select one idle substitute from the $n - 1$ mutual neighbors of r in the dimension e_2, and so on. In such a way, we can maximize the reuse gain of the existing Incast tree and thus significantly reduce the computation overhead for generating a new Incast tree after updating its receiver. Moreover, the intermediate nodes may already cache a lot of data in the original Incast minimal tree; all these data can be able to be reused in the new Incast minimal tree.

7.4 Discussion

In this section, we further discuss some other design factors that affect the aggregation methods.

7.4.1 Extension to General Incast Transfers

To ease the presentation, we focus on an Incast transfer in BCube(n, k), where the maximum hamming distance between the receiver and each sender is $k + 1$. Our RS-based and ARS-based approaches for building an Incast tree can be easily extended to involve general Incast transfers. Let d denote the maximum hamming distance between the receiver and each sender in an Incast transfer. If $d < k + 1$, there exist $k + 1 - d$ dimensions in each of which the labels of all Incast members are identical. A desired Incast tree of the Incast transfer should be $d + 1$ stages with the receiver.

In such a case, Definition 7.3 can be revised as follows. Let $e_1 e_2 \ldots e_j \ldots e_d$ denote a routing sequence with d routing symbols. It is one of $d!$ permutations of d dimensions in each of which the labels of all Incast members are not identical. The stage j of an Incast tree is associated with a routing symbol e_j for $1 \le j \le d$. So far, our RS-based and ARS-based building methods can be directly utilized to calculate an Incast tree for the general Incast transfer.

7.4.2   Extension to Other Network Topologies

As discussed above, for existing switch-centric network topologies, a traditional switch usually does not deliver programmable data plane; hence, they cannot naturally support the inter-flow data aggregation. Fortunately, new Cisco ASIC and Arista application switches have provided a programmable data plane. If future data centers utilize such kind of novel switches, they can perform inter-flow data aggregation in switch-centric structures.

Although we use BCube as a vehicle to study the Incast tree building methods for server-centric network topologies, the proposed methods can be applied to other server-centric network topologies, such as DCell [13], BCN [16], FiConn [27], SWDC [28], and Scafida [18]. The Incast-tree building method, however, has to exploit the topological features of different network topologies.

Note that the proposed methodologies can also be applied to FBFLY and HyperX, two switch-centric network topologies. The main reason is that the topology behind BCube is an emulation of the generalized hypercube while the topologies of FBFLY and HyperX at the level of switch are just the generalized hypercube. For FBFLY and HyperX, our building methods can be used directly if such topologies utilize those novel switches each with a programmable data plane.

7.4.3   Impact on the Job Execution Time

Given a Mapreduce-like job, its execution time depends on three stages, i.e., map, Shuffle, and reduce. The Shuffle duration is dominated by the amount of delivered network traffic and the network resource it can utilize. The introduction of in-network aggregation just imposes impact on the durations of Shuffle and reduce stages, while has no impact on the map stage.

Distributed computing frameworks like MapReduce suffer the map-skew problem in data centers. It means that the map tasks exhibit highly variable task runtimes due to the imbalanced load among them. When such a skew happens, some map tasks take longer to process their input data than others, slowing down the entire computation of a job. Similar results hold in the setting of reduce tasks. Recently, many approaches have been proposed to tackle or mitigate the map-skew problem [29–31]. Such orthogonal approaches can be adopted to support the inter-flow aggregation by completing all of map tasks as simultaneous as possible.

In such settings, each sender of the Incast transfer greedily delivers a data flow toward the receiver along the Incast tree if its output for the receiver has been generated and it is not an aggregating server. When meets an aggregating server, all packets of a flow will be cached. Upon the data from all its children and itself (if any) have been received, the aggregating server aggregate such flows into a new data flow. Thus, our method decreases the Shuffle time. The cause is that our method directly lowers down the network traffic during the Shuffle stage in data centers, hence shortening the Shuffle duration.

Additionally, data aggregation can be performed at an opportunistic level, due to the limits on RAM at aggregating nodes and a few straggler map tasks. That is, an aggregating server performs the aggregation operation once a new data flow arrives. Such a scheme amortizes the delay due to wait and simultaneously aggregate all incoming flows. In a MapReduce cluster of Facebook with 600 nodes, the statistical analysis found that 83% of jobs have less than 271 Map tasks and 30 Reduce tasks [32]. Literature [33, 34] verified that in Hadoop cluster of Facebook and Dryad cluster of Bing, the input data sizes satisfy the long-tail distribution, and the task sizes (input data size and task amount) follow the power-law distribution.

Our inter-flow data aggregation methods are well suitable to small jobs, since each aggregating server is sufficient to cache data packets. In addition, for most MapReduce jobs, functions running on each aggregating server are correlated and interchangeable. In literature [35], authors proposed the method to transform the noninterchangeable and uncorrelated functions, and customized functions into correlated and interchangeable functions. Thus, each aggregating server can operate aggregation function once it receives partial data packets.

7.5   Performance Evaluation

We start with our prototype implementation. We then evaluate our methods and compare with other works under different data center sizes, Incast transfer sizes, aggregation ratios, and distributions of Incast members.

7.5.1   The Prototype Implementation

Our prototype consists of 81 virtual machines (VMs) hosted by 8 physical servers connected together with a switched Ethernet. Each server is equipped with two 8-core hyper-threaded Intel Xeon E5620 2.40 GHz processors, 24 GB memory and a 1 TB

SATA disk, running an unmodified version of CentOS 5.6 with kernel version 2.6.18. Seven of the servers run 10 virtual machines as the Hadoop virtual slave nodes, and the other one runs 10 virtual slave nodes and 1 master node. Each virtual slave node supports four map tasks and one reduce task. All VMs on a physical server share the hosts' NICs through a virtual switch. We extend the Hadoop implementation to embrace the in-network packet cache and the inter-flow data aggregation.

To achieve an Incast transfer, we launch the built-in wordcount job in Hadoop 0.21.0. The number of senders (map tasks) and receivers (reduce tasks) for such a job are set to 320 and 1, respectively. Each Map task is assigned with ten input files, each of which is 64 MB. In the Shuffle phase of such a job, the average amount of data from a sender to the receiver is about 1 MB after performing the combiner at each sender.

We aim to deploy such an Incast transfer in BCube$(6, k)$ data centers for $3 \leq k \leq 9$. To do so, we associate each sender and the only receiver with a $k + 1$ dimensional BCube label in a random way. Conceptually, we achieve a random deployment of the Incast transfer in BCube data centers and can generate an ARS-based Incast tree and a unicast-driven Incast tree.

To deploy each of the resultant Incast tree for the Incast transfer into BCube-based data centers, we design an injection mapping from all vertices in the Incast tree to our testbed as follows. The master VM acts as the Incast manager. Four senders, one possible receiver, and some inner vertices of the Incast tree are mapped to each of the 30 slave VMs. In such a way, we abstract each VM as multiple VMAs (virtual machine agent) each of which corresponds to a vertex in the Incast tree. We require that all vertices mapped to the same VM should not contain any pair of neighbors across successive stages in the Incast tree. For example, vertices v_5 and v_9 in Fig. 7.2c can appear at an identical VM that, however, cannot accommodate the vertex v_1. Thus, no local communication will happen between VMAs inside a VM when performing the Incast transfer. Actually, for any VMA its neighbor at the next stage in the Incast tree appears at different VMs on the same even different physical servers. Thus, each edge in the Incast tree is mapped to a virtual link between two VMs or a physical link across two servers in our testbed.

We then compare our ARS-based Incast tree against the typical Steiner-tree algorithm, the unicast-driven Incast tree, and the existing method in terms of four metrics. They are the resultant network traffic, the number of active links, the number of cache servers, and the input data size at the receiver. The network traffic denotes the sum of network traffic over all edges in the Incast tree.

The Steiner-tree algorithm we choose is the one described in literature [36], whose benefit is the computation speed. The algorithm works as follows: (1) A virtual complete graph is generated upon the Incast members; (2) A minimum spanning tree is calculated on the virtual complete graph; (3) The virtual link in the virtual complete graph is replaced by the shortest path between any two Incast members in the original topology, with unnecessary links deleted.

7.5.2 Impact of the Data Center Size

Conceptually, we deploy an Incast transfer with 320 senders over a subnet of BCube(6,k) on our testbed given the resultant Incast tree. We conduct experiments and then collect the performance metrics after completing such an Incast transfer along the Incast tree. We also carry out similar experiments for the same Incast transfer under the Steiner tree algorithm, the unicast-driven Incast tree and the Incast tree resulting from the existing method. Figure 7.3 shows the changing trends of the performance metrics on average, among 1000 rounds of experiments, under different methods and settings.

Figure 7.3a suggests that the ARS-based Incast tree, and the unicast-driven tree can save the average resultant network traffic by 38% and 18%, respectively, compared with the existing method. The reason is that the number of aggregating servers in the former one increases while that in the latter ones decreases along with the increase of k. Additionally, ARS-based Incast tree utilizes less links, and hence less servers and

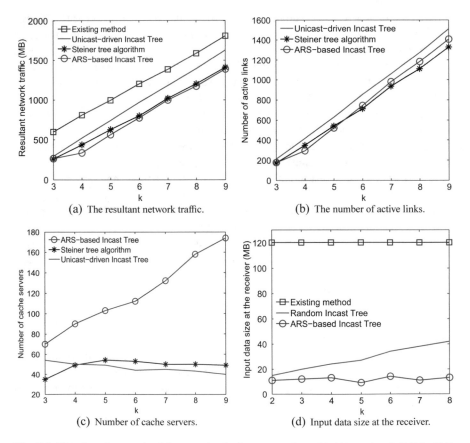

(a) The resultant network traffic.

(b) The number of active links.

(c) Number of cache servers.

(d) Input data size at the receiver.

Fig. 7.3 The changing trends of four metrics for incast transfers with 320 senders in BCube(6,k)

Table 7.1 Impact of the data center size on the delay during the Shuffle and reduce phases (s)

BCube(6, k)	k = 4	k = 5	k = 6	k = 7	k = 8	k = 9
ARS-based approach	211	231	265	292	318	335
Existing method	937	1109	1120	1264	1312	1343

network devices. Moreover, ARS-based approach significantly reduces the amount of received data at the receiver, as shown in Fig. 7.3d. Such benefits of in-network aggregation considerably reduce the total delay during the Shuffle and reduce phases of a job, as shown in Table 7.1. The delay during the map phase is the same for all methods.

In addition, ARS-based approach outperforms Steiner tree algorithm to some extent, because the former one utilizes more aggregation servers than the latter one when they have the same number of links. In summary, ARS-based approach always achieves far better performance than the other methods, irrespective of the size of a data center.

7.5.3 Impact of the Incast Transfer Size

In the current practice, a MapReduce-like job often involves several hundred even thousands map tasks. Our testbed, however, cannot perform a large scale job due to its limited resources. We thus conduct simulations to demonstrate the scaling properties of our method. We achieves BCube network structure with Java, and generate Incast transfer according to the wordcount job in Hadoop. Specifically, we create a set of Incast transfers with m senders for $m \in \{100, 200, \ldots, 3900, 4000\}$. The wordcount job provides 64 MB input data for each sender, randomly generated by the built-in RandomTextWriter of Hadoop based on a 1000-word dictionary. The data transmission from each sender to the receiver is controlled to be 1 MB on average. Figure 7.4 shows the network traffic and the number of active links on average under varying sizes of Incast transfers in BCube (8,5). The number of servers accommodated by a data center with BCube(8,5) is 262,144, which is large enough for a production data center.

Results in Fig. 7.4a indicate that both our ARS-based and the unicast-driven Incast trees considerably save the network traffic by 59 and 27% on average compared to the existing method, as m ranges from 100 to 4000. Such results demonstrate the gain of the inter-flow data aggregation even in large-scale Incast transfers. Figure 7.4b also indicates that the ARS-based Incast tree utilizes less links (hence less servers and network devices) than the unicast-driven Incast tree.

In summary, ARS-based approach can support large-scale Incast transfer well, and it outperforms other methods in terms of network traffic and resource utilization.

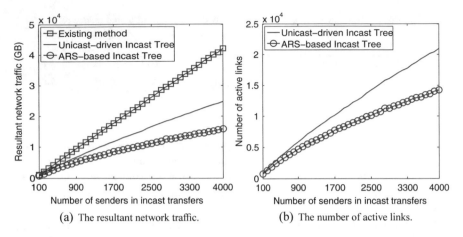

(a) The resultant network traffic. (b) The number of active links.

Fig. 7.4 The changing trends of two metrics along with the increase of senders in an incast transfer in BCube(8,5)

Table 7.2 Impact of the incast transfer size on the delay during the shuffle and reduce phases (seconds) in Bcube(8,5)

	500	1000	1500	2000	2500	3000
m						
ARS-based approach	269	282	293	300	302	310
Existing method	858	1241	1637	2107	2630	3215

Moreover, ARS-based approach incurs less delay, during the Shuffle and reduce phases of an Incast transfer, compared to the existing method, as shown in Table 7.2.

7.5.4   Impact of the Aggregation Ratio

We assume earlier that data flows consisting of key-value pairs can be aggregated as a new one whose size is equal to that of the largest among all participants. In other words, the set of keys in each involved data flow is the subset of that in the largest data flow. We further evaluate the performance of our ARS-based approach in a general Incast transfer.

Given β data flows in a general Incast transfer, let c_i denote the size of the ith data flow for $1 \leq i \leq \beta$. Let α denote the aggregation ratio among any number of data flows, where $0 \leq \alpha \leq 1$. Thus, the size of the resultant new data flow after aggregating such β ones is given by

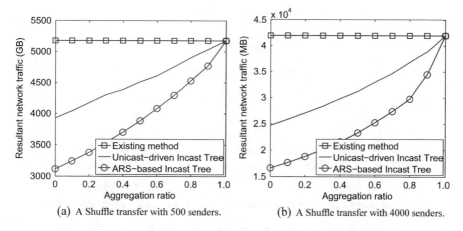

(a) A Shuffle transfer with 500 senders. (b) A Shuffle transfer with 4000 senders.

Fig. 7.5 The resultant network traffic under different aggregation ratios BCube(8,5)

$$\max\{f_1, f_2, \ldots, f_s\} + \delta \times \left(\sum_{i=1}^{s} f_i - \max\{f_1, f_2, \ldots, f_s\}\right).$$

It is clear that our analysis in Sect. 8.2 and the simulations for large-scale Incast in this section just tackle a special scenario where $\delta = 0$. When $\delta = 1$, each key in one of such s data flows does not appear at other data flows, thus, the data aggregation among the s data flows does not bring any gain. Such two special scenarios rarely happen in practice. Therefore, we conduct extensive simulations to evaluate ARS-based approach in a more general scenario where $0 \le \delta \le 1$.

The amount of network traffic generated by ARS-based approach, unicast-driven approach, and the existing methods are evaluated in BCube(8,5) when δ varies from 0 to 1. Figure 7.5 indicates that the ARS-based approach always incurs much less network traffic than the other two methods, especially for small values of δ. Assume that the value of the random variable δ follows a uniform distribution. In such a setting, ARS-based Incast tree saves the resultant network traffic by 24 and 40%, compared to the existing method, when the Incast transfer involves 500 and 4000 senders, respectively.

Figure 7.6 reports the delay during the Shuffle and reduce stages as we vary the aggregation factor δ in BCube(8,5). For each of the two Shuffle transfers, ARS-based approach always achieves the lower Shuffle and reduce time across all values of aggregation ratio except that the aggregation ratio is 1. In such an extreme case, our method is equivalent to the existing method since the data aggregation will not reduce the amount of network traffic.

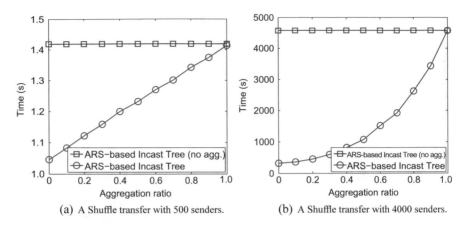

(a) A Shuffle transfer with 500 senders.   (b) A Shuffle transfer with 4000 senders.

Fig. 7.6 The shuffle and reduce time under different aggregation ratios in BCube(8,5)

7.5.5 Impact of the Distribution of Incast Members

In the state-of-the-practice, all members of an Incast transfer are scheduled to idle servers in the scope of the entire data center. In such a case, the senders and the receiver are somehow randomly distributed in the data center. Such a random distribution makes the Incast transfer occupy more data center resources and generate more network traffic. We further study the impact of different member distribution models on the gain of inter-flow aggregation.

The key idea is to find the smallest subnet BCube(n, k_1) in BCube(n, k) such that all members of an Incast transfer satisfy their schedule constraints in the subnet. In such a case, the hamming distance between the receiver and each sender of the Incast transfer is at most $k_1 + 1 \leq k + 1$. The ARS-based Incast tree in the scope of BCube(n, k_1) for such an Incast transfer has at most $k_1 + 1$ stages. In theory, it thus occupies less data center resources and incurs less network traffic than prior ARS-based Incast tree in the scope of BCube(n, k).

We generate 30 Shuffle transfers each of which has 500 senders but different number of receivers, ranging from 1 to 30. For each Shuffle transfer, we apply the two distribution schemes, i.e., managed distribution and random distribution, to all its Incast transfers in BCube(8, 5) and calculate the cumulative network traffic. Figure 7.7 indicates that both ARS-based approach and the existing method result in less network traffic in the managed distribution scheme than that in the random distribution scheme. Moreover, ARS-based approach saves the network traffic by 62 and 24% on average compared to the existing method in the managed and random distribution schemes, respectively. Such results indicate that ARS-based approach can achieve more gains in the managed distribution scheme.

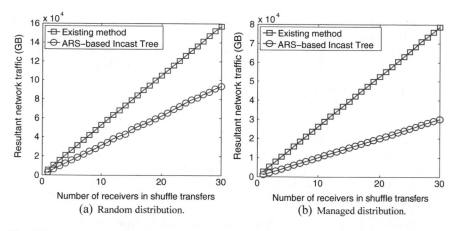

Fig. 7.7 The network traffic under different distributions of shuffle member in BCube(8,5)

References

1. Condie T, Conway N, Alvaro P, et al. MapReduce Online [C]. In Proc. of 7th USENIX NSDI, San Jose, USA, 2010: 313–328.
2. Yu Y, Isard M, Fetterly D, et al. DryadLINQ: A System for General-Purpose Distributed Data-Parallel Computing Using a High-Level Language [C]. In Proc. of 8th USENIX OSDI, San Diego, USA, 2008: 1–14.
3. Murray D G, Schwarzkopf M, Smowton C, et al. CIEL: a universal execution engine for distributed data-flow computing [C]. In Proc. of 8th USENIX NSDI, Boston, USA, 2011.
4. Gueron M, Llia R, Margulis. Pregel: a system for large-scale graph processing [C]. In Proc. of ACM SPAA, Calgary, Alberta, Canada, 2009: 135–146.
5. Zaharia M, Chowdhury M, Franklin M J, et al. Spark: Cluster computing with working sets [J]. Book of Extremes, 2010, 15(1): 1765–1773.
6. Chowdhury M, Zaharia M, Ma J, et al. Managing data transfers in computer clusters with orchestra [C]. In Proc. of ACM SIGCOMM, Athens, Greece, 2011: 98–109.
7. Al-Fares A, Loukissas A, Vahdat A. A scalable, commodity data center network architecture [C]. In Proc. of ACM SIGCOMM, Seattle, USA, 2008: 63–74.
8. Greenberg A, Jain N, Kandula S, et al. VL2: A scalable and flexible data center network [C]. In Proc. of ACM SIGCOMM, Barcelona, Spain, 2009: 51–62.
9. Mysore R, Pamboris A, Farrington N. PortLand: A scalable fault-tolerant layer 2 data center network fabric [C]. In Proc. of ACM SIGCOMM, Barcelona, Spain, 2009: 39–50.
10. Abts D, Marty M A, Wells P M, et al. Energy proportional datacenter networks [C]. In Proc. of ACM ISCA, Saint-Malo, France, 2010: 338–347.
11. Ahn J H, Binkert N L, Davis A, et al. HyperX: topology, routing, and packaging of efficient large-scale networks [C]. In Proc. of IEEE/ACM SC, Santa Clara, USA, 2009: 1–11.
12. Singla A, Hong C Y, Popa L, et al. Jellyfish: Networking data centers randomly [C]. In Proc. of 9th USENIX NSDI, San Jose, USA, 2012: 225–238.
13. Guo C, Wu H, Tan K, et al. DCell: A scalable and fault-tolerant network structure for data centers [C]. In Proc. of ACM SIGCOMM, Seattle, USA, 2008: 75–86.
14. Guo C, Lu G, Li D, et al. BCube: A high performance, server-centric network architecture for modular data centers [J]. ACM SIGCOMM Computer Communication Review, 2009, 39(4): 63–74.
15. Abu-Libdeh H, Costa P, Rowstron A, et al. Symbiotic routing in future data centers [J]. ACM SIGCOMM Computer Communication Review, 2010, 40(4): 51–62.

16. Guo D, Chen T, Li D, et al. BCN: Expansible network structures for data centers using hierarchical compound graphs [C]. In. Proc. of IEEE INFOCOM, Shanghai, China, 2011: 61–65.
17. Guo D, Chen T, Li D, et al. Expansible and cost-effective network structures for data centers using dual-port servers [J]. IEEE Transactions on Computers, 2012, 62(7): 1303–1317.
18. Gyarmati L, Trinh T. Scafida: A scale-free network inspired data center architecture [J]. ACM SIGCOMM Computer Communication Review, 2010, 40(5): 4–12.
19. Condie T, Conway N, Alvaro P, et al. Online aggregation and continuous query support in MapReduce [C]. In Proc of ACM SIGMOD, Indianapolis, USA, 2010: 1115–1118.
20. Pansare N, Borkar V R, Jermaine C, et al. Online aggregation for large MapReduce jobs [J]. Proceeding of the Vldb Endowment, 2011, 4(11): 1135–1145.
21. Lu G, Guo C, Li Y, et al. ServerSwitch: A programmable and high performance platform for data center networks [C]. In Proc. of USENIX NSDI, Boston, USA, 2011: 15–28.
22. Cao J, Guo C, Lu G, et al. Datacast: a scalable and efficient reliable group data delivery service for data centers [J]. IEEE Journal on Selected Areas in Communications, 2012, 31(31): 2632–2645.
23. Li D, Li Y, Wu J, et al. ESM: Efficient and scalabledata center multicast routing [J]. IEEE/ACM Transactions on Networking, 2012, 20(3): 944–955.
24. Hong C Y, Kandula S, Mahajan R, et al. Achieving high utilization with software-driven WAN[J]. ACM SIGCOMM Computer Communication Review, 2013, 43(4): 15–26.
25. Yeganeh S H, Tootoonchian A, Ganjali Y. On scalability of software-defined networking [J]. IEEE Communications Magazine, 2013, 51(2): 136–141.
26. Bhuyan L N, Agrawal D P. Generalized hypercube and hyperbus structures for a computer network [J]. IEEE Transactions on Computers, 1984, 100(4): 323–333.
27. Li D, Guo C, Wu H, et al. Scalable and cost-effective interconnection of data-center servers using dual server ports [J]. IEEE/ACM Transactions on Networking, 2011, 19(1): 102–114.
28. Shin J Y, Wong B, Sirer E G. Small-world datacenters [C]. In Proc. of 2nd ACM SOCC, Cascais, Portugal, 2011: 2–14.
29. Kwon Y C, Balazinska M, Home B, et al. SkewTune: Mitigating skew in MapReduce applications [C]. In Proc. of ACM SIGMOD, Scottsdale, USA, 2012: 25–36.
30. Kwon Y, Balazinska M, Howe B, et al. Skew-resistant parallel processing of feature-extracting scientific user-defined functions [C]. In Proc. of 1st ACM SOCC, Indianapolis, USA, 2010: 75–86.
31. Kwon Y, Ren K, Balazinska M, et al. Managing skew in Hadoop [J]. IEEE Data Engineering Bulletin, 2013, 36(1): 24–33.
32. Zaharia M, Borthakur D, Sarma J S, et al. Delay scheduling: a simple technique for achieving locality and fairness in cluster scheduling [C]. In Proc. of 5th ACM EuroSys, Paris, France, 2010: 265–278.
33. Ananthanarayanan G, Ghodsi A, Wang A, et al. PACMan: Coordinated memory caching for parallel jobs [C]. In Proc. of USENIX NSDI, San Jose, USA, 2012: 267–280.
34. Chen Y, Alspaugh S, Borthakur D, et al. Energy efficiency for large-scale mapreduce workloads with significant interactive analysis [C]. In Proc. of 7th ACM EuroSys, Bern, Switzerland, 2012: 43–56.
35. Yu Y, Gunda P K, Isard M. Distributed aggregation for data-parallel computing: Interfaces and implementations [C]. In Proc. of 22nd ACM SOSP, Big Sky, USA, 2009: 247–260.
36. Kou L, Markowsky G, Berman L. A fast algorithm for Steiner trees [J]. Acta Informatica (Historical Archive), 1981, 15(2):141–145.

Chapter 8
Collaborative Management of Correlated Shuffle Transfer

Abstract The inter-flow aggregation of correlated Incast transfer is introduced in Chap. 7. This chapter considers how to realize the in-network aggregation of correlated Shuffle transfer, so that the consumed network resources can be considerably reduced. Specifically, the in-network aggregation problem of Shuffle transfer is formalized for a BCube data center. To tackle this NP-hard problem, we propose two approximate methods for efficiently constructing the shuffle aggregation subgraph, solely based on the labels of their members and the data center topology. The expected in-network aggregation can be effectively achieved through the collaborative transmission of traffic based on this subgraph. This chapter also introduces the scalable traffic forwarding mode based on Bloom filters, so as to achieve the desired in-network aggregation effect for a large number of coexisting Shuffle transfers. Although this chapter chooses BCube as the network topology, the concept of Shuffle in-network aggregation is applicable to other types of data center topologies.

8.1 Introduction

Distributed computing systems like MapReduce [1] in data centers transfer massive amount of data across successive processing stages. Such Shuffle transfers contribute most of the network traffic and make the network bandwidth become a bottleneck. To improve the network performance of data centers, academia and Industry proposed many novel network interconnection structures, such as BCube [2]. However, it is more important to efficiently use the available network bandwidth of data centers, compared to just increasing the network capacity.

Similar to Chap. 7, this chapter focuses on managing the network activity at the level of transfers so as to significantly lower down the network traffic and efficiently use the available network bandwidth. The many-to-many Shuffle and many-to-one Incast transfers are the two most common transfer patterns, and contribute most of the network traffic (about 80% as shown in [3]) and impose severe impacts on the application performance. As described in Chap. 7, the data flows from all senders to each receiver in a Shuffle transfer are typically highly correlated. For example, each reducer of a MapReduce job is assigned a unique partition of the key range

© Springer Nature Singapore Pte Ltd. 2022
D. Guo, *Data Center Networking*,
https://doi.org/10.1007/978-981-16-9368-7_8

and performs aggregation operations on the content of its partition retrieved from every mapper's output. Such aggregation operations can be the SUM, MAX, MIN, COUNT, TOP-K, KNN, et al. As prior work has shown, the reduction in the size between the input data and output data of the receiver after aggregation is 81.7% for Mapreduce jobs in Facebook [4].

To reduce the network traffic and efficiently use the available network bandwidth, this chapter will push the aggregation computation into the network rather than just at the receiver side. In this chapter, we examine the gain and feasibility of the in-network aggregation on a Shuffle transfer in a server-centric data center, and formalize the in-network aggregation of Shuffle transfers. To maximize the gain of in-network aggregation of Shuffle transfers, which is an NP-hard problem, we propose two approximate methods, called IRS-based and SRS-based methods. We further design scalable forwarding schemes based on Bloom filters to implement in-network aggregation on massive concurrent Shuffle transfers.

We prove that the in-network aggregation of Shuffle transfers can effectively reduce the network traffic and save the network resources with a prototype implementation and large-scale simulations. Actually, the SRS-based approach saves the network traffic by 32.87% on average for a small-scale Shuffle transfer with 120 members in data centers BCube$(6, k)(2 \leq k \leq 8)$. It saves the network traffic by 55.33% on average for Shuffle transfers with 100–3000 members in a large-scale data center BCube$(8, 5)$ with 262,144 servers. Although these methods are proposed based on BCube, they can be applied to other server-centric structures after simple modifications.

8.2   In-Network Aggregation of Shuffle Transfers

8.2.1   Problem Statement

We model a data center as a graph $G = (V, E)$ with a vertex set V and an edge set E. A vertex of the graph refers a switch or a datacenter server. An edge (u, v) denotes a link, through which u connects with v where $v, u \in V$.

Definition 8.1 A Shuffle transfer has m senders and n receivers where a data flow is established for any pair of sender i and receiver j for $1 \leq i \leq m$ and $1 \leq j \leq n$. An Incast transfer consists of m senders and one of n receivers. A Shuffle transfer consists of n Incast transfers that share the same set of senders but differ in the receiver.

In many commonly used workloads, data flows from all of m senders to all of n receivers in a Shuffle transfer are highly correlated. More precisely, for each of its n Incast transfers, the list of key-value pairs among m flows share the identical key partition for the same receiver. For this reason, the receiver typically applies an aggregation function, e.g., the SUM, MAX, MIN, COUNT, TOP-K, and KNN, to the received data flows in an Incast transfer. As prior work has shown, the reduction in the size between the input data and output data of the receiver after aggregation is 81.7% for Facebook jobs [4].

In this chapter, we aim to push aggregation into the network and parallelize the Shuffle and reduce phases so as to minimize the resultant traffic of a Shuffle transfer significantly. We start with a simplified Shuffle transfer with $n = 1$, an Incast transfer, and then discuss a general Shuffle transfer. Given an Incast transfer with a receiver r and m senders, message routes from senders to the same receiver essentially form an aggregation tree. In principle, there exist many such trees for an Incast transfer in densely connected data center networks, e.g., BCube. Although any tree topology could be used, tree topologies differ in their aggregation gains. A challenge is how to produce an aggregation tree that minimizes the amount of network traffic of a Shuffle transfer after applying the in-network aggregation.

Given an aggregation tree in BCube, we define the total edge weight as its cost metric, i.e., the sum of the amount of the outgoing traffics of all vertices in the tree. BCube utilizes traditional switches and only data center servers enable the in-network caching and processing. Thus, a vertex in the aggregation tree is an aggregating vertex only if it represents a server and at least two flows converge at it. An aggregating vertex then aggregates its incoming flows and forwards a resultant single flow instead of multiple individual flows along the tree. At a non-aggregating vertex, the size of its outgoing flow is the cumulative size of its incoming flows. The vertices representing switches are always non-aggregating ones. We assume that the introduced traffic from each of m senders is 1MB so as to normalize the cost of an aggregation tree. In such a way, the weight of the outgoing link is one at an aggregating vertex and equals to the number of incoming flows at a non-aggregating vertex.

Definition 8.2 For an Incast transfer, the minimal aggregation tree problem is to find a connected subgraph in $G = (V, E)$ that spans all Incast members with minimal cost for completing the Incast transfer. As proved in Chap. 7, the minimal aggregation tree of an Incast transfer in BCube is NP-hard.

Theorem 8.1 *Given any Incast transfer in BCube, finding the minimum aggregation tree is NP-hard.*

The problem close to the minimum aggregation tree of Incast transfer is the Steiner minimum tree (SMT) problem of multicast transfer. There are many approximate algorithms for Steiner tree problem. The time complexity of such algorithms for general graphs is of $O(m \times N^2)$, where m and N are the numbers of Incast numbers and all servers in a data center. The time complexity is too high to meet the requirement of online tree building for Incast transfers in production data centers which hosts large number of servers. On the other hand, such algorithms cannot efficiently exploit the topological feature of BCube, a densely connected data center network. For such reasons, we develop an efficient Incast tree building method by exploiting the topological feature of BCube in Sect. 8.2.2, and the resulting time complexity is of $O(m \times (\log N)^3)$.

Definition 8.3 For a Shuffle transfer, the minimal aggregation subgraph problem is to find a connected subgraph in $G = (V, E)$ that spans all Shuffle members with minimal cost for completing the Shuffle transfer.

Note that the minimal aggregation tree of an Incast transfer in BCube is NP-hard and a Shuffle transfer is normalized as the combination of a set of Incast transfers. Therefore, the minimal aggregation subgraph of a Shuffle transfer in BCube is NP-hard.

After deriving the Shuffle subgraph for a given Shuffle, each sender greedily delivers its data flow toward a receiver along the Shuffle subgraph if its output for the receiver is ready. All packets will be cached when a data flow meets an aggregating server. An aggregating server can perform the aggregation operation once a new data flow arrives and brings an additional delay. However, the in-network aggregation can significantly reduce network traffic, and share the delay in Reduce stage among aggregating servers. Thus, the job completion time can be decreased, and this will be proved in experiment sections.

8.2.2   Construction of an Incast Aggregation Tree

Actually, we have introduced two approximate building methods of minimal cost aggregation tree for Incast transfer in Chap. 7. Considering the new characteristics of Shuffle, this chapter further proposes a novel approximate method of minimal cost Incast aggregation tree, namely, IRS-based method. This method exploits the topological feature of BCube(n, k). For ease of understanding, we describe the construction process of this method in detail.

In densely connected network structures, many equal-cost disjoint paths can be found between any server pair. For example, in BCube(n, k), there are $k + 1$ equal-cost disjoint Unicast paths between any server pair if their labels differ in $k + 1$ dimensions. For an Incast transfer, each of its all senders independently delivers its data to the receiver along a Unicast path that is randomly selected from $k + 1$ ones. The combination of such Unicast paths forms a Unicast-based aggregation tree, as shown in Fig. 8.1a. Such a method, however, has less chance of achieving the gain of the in-network aggregation.

For an Incast transfer, we aim to build an efficient aggregation tree in a managed way instead of the random way, given the labels of all Incast members and the data center topology. Consider an Incast transfer consists of a receiver and m senders in BCube(n, k). Let d denote the maximum hamming distance between the receiver and each sender in the Incast transfer, where $d \leq k + 1$. Without loss of generality, we consider the general case that $d = k + 1$. Like the case in Chap. 7, an aggregation tree of the Incast transfer can be expanded as a directed multistage graph with $d + 1$ stages. The stage 0 only has the receiver. Each of all senders should appear at stage j if it is a j-hop neighbor of the receiver. Only such senders and the receiver, however, cannot definitely form a connected subgraph. The problem is then translated to identify a minimal set of servers for each stage to ensure the resulted subgraph connected. The level and label of a switch between a pair of neighboring servers across two stages can be inferred from the labels of the two servers. We thus only focus on identify additional servers for each stage, without considering the switch selections.

Identifying the minimal set of servers for each stage is an efficient approximate method for the minimal aggregation tree problem. For any stage, less number of servers incurs less number of flows towards the receiver since all incoming flows at each server will be aggregated as a single flow. Given the server set at stage j for $1 \leq j \leq d$, we can infer the required servers at stage $j - 1$ by leveraging the topology features of BCube(n, k). Each server at stage j has j one-hop neighbors at stage $j - 1$. If each server at stage j randomly selects one of j one-hop neighbors at stage $j - 1$, it just results in a Unicast-based aggregation tree.

The insight of our method is to derive a common neighbor at stage $j - 1$ for as many servers at stage j as possible. In such a way, the number of servers at stage $j - 1$ can be significantly reduced and each of them can merge all its incoming flows as a single one for further forwarding. We identify the minimal set of servers for each stage from stage d to stage 1 along the processes as follows.

We start with stage $j = d$ where all servers are just those senders that are d-hops neighbors of the receiver in the Incast transfer. After defining a routing symbol $e_j \in \{0, 1, ..., k\}$, we partition all such servers into groups such that all servers in each group are mutual one-hop neighbors in dimension e_j. That is, the labels of all servers in each group only differ in dimension e_j. In each group, all servers establish paths to a common neighboring server at stage $j - 1$ that aggregates the flows from all servers in the group, called an *inter-stage aggregation*. All servers in the group and their common neighboring server at stage $j - 1$ have the same label except the e_j dimension. Actually, the e_j dimension of the common neighboring server is just equal to the receiver's label in dimension e_j. Thus, the number of partitioned groups at stage j is just equal to the number of appended servers at stage $j - 1$. The union of such appended servers and the possible senders at stage $j - 1$ constitute the set of all servers at stage $j - 1$. For example, all servers at stage 2 are partitioned into three groups, $\{v_5, v_9\}$, $\{v_{10}, v_{14}\}$, and $\{v_{11}\}$, using a routing symbol $e_2 = 1$, as shown in Fig. 8.1b. The neighboring servers at stage 1 of such groups are with labels v_1, v_2, and v_3, respectively.

On this basis, the number of servers at stage $j - 1$ still has an opportunity to be reduced due to the observations as follows. There may exist groups each of which only has one element at each stage, and has no chance to aggregate flows. For example, in Fig. 8.1b, the group $\{v_{11}\}$ only contains one server, and flows from it to the receiver has no opportunity to perform the in-network aggregation.

To address such an issue, we propose an *intra-stage aggregation* scheme. After partitioning all servers at any stage j according to e_j, we focus on groups each of which has a single server and its neighboring server at stage $j - 1$ is not a sender of the Incast transfer. That is, the neighboring server at stage $j - 1$ for each of such groups fails to perform the inter-stage aggregation. At stage j, the only server in such a group has no one-hop neighbors in dimension e_j, but may have one-hop neighbors

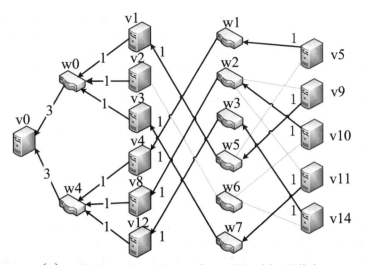

(a) A Unicast-based tree of cost 22 with 18 links.

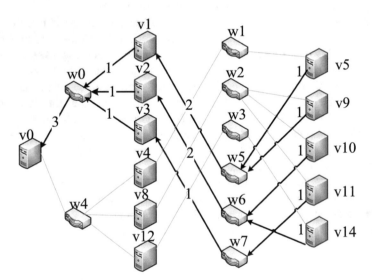

(b) An aggregation tree of cost 16 with 12 links.

Fig. 8.1 Different aggregation trees for an incast transfer with the sender set $\{v_2, v_5, v_9, v_{10}, v_{11}, v_{14}\}$ and the receiver v_0 in Bcube(4,1)

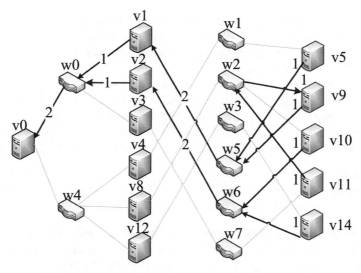

(c) An aggregation tree of cost 14 with 11 links.

Fig. 8.1 (continued)

in other dimensions, $\{0, 1, ..., k\} - \{e_j\}$. In such a case, the only server in such a group no longer delivers its data to its neighboring server at stage $j - 1$ but to a one-hop neighbor in another dimension at stage j. The selected one-hop neighbor thus aggregates all incoming data flows and the data flow by itself (if any) at stage j, called an *intra-stage aggregation*.

Such an intra-stage aggregation can further reduce the cost of the aggregation tree for groups in stage j each of which has a single server. The cost of the one-hop forwarding is 2 due to the existence of switches, no matter in inter-stage forwarding or intra-stage forwarding. However, introducing such a one-hop forwarding saves the cost by at least four in inter-stage forwarding. The root cause is that for a sole server in a group at a stage j the nearest aggregating server along the original path to the receiver appears no earlier than stage $\leq j - 2$. For example, the server v_{11} has no neighbor in dimension $e_2 = 1$ at stage 2, however, has two neighbors, the servers v_9 and v_{10}, in dimension 0 at stage 2, as shown in Fig. 8.1b. If the server v_{11} uses the server v_9 or v_{10} as a relay for its data flow towards the receiver, prior aggregation tree in Fig. 8.1b can be optimized as the new one in Fig. 8.1c. Thus, the cost of the aggregation tree is decreased by two while the number of active links is decremented by one.

So far, the set of servers at stage $j - 1$ can be achieved under a partition of all servers at stage j according to dimension e_j. The number of outgoing data flows from stage $j - 1$ to its next stage $j - 2$ is just equal to the cardinality of the server set at stage $j - 1$. Inspired by such a fact, we apply the above method to other k settings of e_j and accordingly achieve other k server sets that can appear at the stage

$j - 1$. We then simply select the smallest server set from all $k + 1$ ones. Thus, the data flows from stage j can achieve the largest gain of the in-network aggregation at the next stage $j - 1$. The setting of e_j is marked as the best routing symbol for stage j among all $k + 1$ candidates.

Similarly, we can further derive the minimal set of servers and the best choice of e_{j-1} at the next stage $j - 2$, given the server set at $j - 1$ stage, and so on. Finally, server sets at all $k + 2$ stages and the directed paths between successive stages constitute an aggregation tree. The behind routing symbol at each stage is simultaneously found. Such a method is called the ***IRS-based aggregation tree building method***.

Theorem 8.2 *Given an Incast transfer consisting of m senders in BCube(n, k), the complexity of the IRS-based aggregation tree building method is $O(m \times (\log N)^3)$, where $N = n^{k+1}$ is the number of servers in a BCube(n, k).*

Proof The IRS-based aggregation tree building method performs at most k stages, from stage $k + 1$ to stage 2. Given an Incast transfer, the process to partition all servers at any stage j into groups according to e_j can be simplified as follows. For each server at stage j, we extract its label except the e_j dimension. The resultant label denotes a group that involves such a server. We then add the server into such a group. The computational overhead of such a partition is proportional to the number of servers at stage j. Thus, the resultant time complexity at each stage is $O(m)$ since the number of servers at each stage cannot exceed m. Additionally, the intra-stage aggregation after a partition of servers at each stage incurs additional $O(k \times m)$ computational overhead.

In the IRS-based building method, we conduct the same operations under all of $k + 1$ settings of e_{k+1}. This produces $k + 1$ server sets for stage k at the cost of $O((k + 1)^2 \times m)$. In summary, the total computation cost of generating the set of servers at stage k is $O((k + 1)^2 \times m)$. At stage k, IRS-based building method can identify the server set for stage $k - 1$ from k candidates resulting from k settings of e_k since e_{k+1} has exclusively selected one from the set $\{0, 1, 2, ..., k\}$. The resultant computation cost is thus $O(k^2 \times m)$. In summary, the total computation cost of the IRS-based building method is $O(k^3 \times m)$ since it involves at most k stages. Thus, Theorem 8.2 proved.   □

8.2.3 Construction of a Shuffle Aggregation Subgraph

For a Shuffle transfer in BCube(n, k), let $S = \{s_1, s_2, ..., s_m\}$ and $R = \{r_1, r_2, ..., r_n\}$ denote the sender set and receiver set, respectively. An intrinsic way for scheduling all flows in the Shuffler transfer is to derive an aggregation tree from all senders to each receiver via the IRS-based building method. The combination of such aggregation trees produces an ***Incast-based Shuffle subgraph*** in $G = (V, E)$ that spans all senders and receivers. The Shuffle subgraph consists of $m \times n$ paths along which all flows in the Shuffle transfer can be successfully delivered. If we produce a Unicast-

---

Algorithm 8.1 Clustering all nodes in a graph $G' = (V', E')$

Require: A graph $G' = (V', E')$ with $|V'| = n$ vertices.
1: Let *groups* denote an empty set;
2: **while** G' is not empty **do**
3:    Calculate the degree of each vertex in V';
4:    Find the vertex with the largest degree in G'. Such a vertex and its neighbor vertices form a group that is added into the set *groups*;
5:    Remove all elements in the resultant group and their related edges from G'.

---

based aggregation tree from all senders to each receiver, the combination of such trees results in a ***Unicast-based Shuffle subgraph***.

We find that the Shuffle subgraph has an opportunity to be optimized due to the observations as follows. There exist $(k + 1) \times (n - 1)$ one-hop neighbors of any server in BCube(n, k). For any receiver $r \in R$, there may exist some receivers in R that are one-hop neighbors of the receiver r. Such a fact motivates us to think whether the aggregation tree rooted at the receiver r can be reused to carry flows for its one-hop neighbors in R. That is, the flows from all of senders to a one-hop neighbor $r_1 \in R$ of r can be delivered to the receiver r along its aggregation tree and then be forwarded to the receiver r_1 in one hop. In such a way, a Shuffle transfer significantly reuses some aggregation trees and thus utilizes less data center resources, including physical links, servers, and switches, compared to the Incast-based method. We formalize such a basic idea as a NP-hard problem in Definition 8.4.

Definition 8.4 For a Shuffle transfer in BCube (n, k), the minimal clustering problem of all receivers is how to partition all receivers into a minimal number of groups under two constraints. (1) The intersection of any two groups is empty; (2) Other receivers are one-hop neighbors of a receiver in each group.

Actually, Definition 8.4 can be relaxed to find a minimal dominating set (MDS) in a graph $G' = (V', E')$, where V' denotes the set of all receivers of the Shuffle transfer, and for any $u, v \in V'$, there is $(u, v) \in E'$ if u and v are one-hop neighbor in BCube(n, k). In such a way, each element of MDS and its neighbors form a group. Any pair of such groups, however, may have common elements. This is the only difference between the minimal dominating set problem and the minimal clustering problem defined in Definition 8.4. Finding a minimum dominating set is NP-hard in general. Therefore, the minimal clustering problem of all receivers for a Shuffle transfer is NP-hard. We thus propose an efficient algorithm to approximate the optimal solution, as shown in Algorithm 8.1.

The basic idea behind this algorithm is to calculate the degree of each vertex in the graph G' and find the vertex with the largest degree. Such a vertex (a head vertex) and its neighbors form the largest group. We then remove all vertices in the group and their related edges from G'. This brings impact on the degree of each remainder vertex in G'. Therefore, we repeat the above processes if the graph G' is not empty. In such a way, Algorithm 8.1 can partition all receivers of any Shuffle transfer as a set of disjoint groups. Let α denote the number of such groups.

We define the cost of a Shuffle transfer from m senders to a group of receivers $R_i = \{r_1, r_2, ..., r_g\}$ as C_i, where $\sum_{I \geq 1}^{\alpha} |R_i| = n$. Without loss of generality, we assume that the group head is r_1 in R_i. Let c_j denote the cost of an aggregation tree, produced by our IRS-based building method, from all m senders to a receiver r_j in the group R_i where $1 \leq j \leq |R_i|$. The cost of such a Shuffle transfer depends on the entry point selection for each receiver group R_i. The details are as follows:

1. $C_i^1 = |R_i| \times c_1 + 2(|R_i| - 1)$ if the group head r_1 is selected as the entry point. In such a case, the flows from all of senders to all of the group members are firstly delivered to r_1 along its aggregation tree and then forwarded to each of other group members in one hop. Such a one-hop forwarding operation increases the cost by two due to the relay of a switch between any two servers.

2. $C_i^j = |R_i| \times c_j + 4(|R_i| - \beta - 1) + 2 \times \beta$ if a receiver r_j is selected as the entry point. In such a case, the flows from all of senders to all of the group members are firstly delivered to r_j along its aggregation tree. The flow towards each of β one-hop neighbors of r_j in R_i, including the head r_1, reaches its destination in one-hop from r_j at the cost of 2. The receiver r_j finally forwards flows towards other $|R_i| - \beta - 1$ receivers, each of which is two-hops from r_j due to the relay of the group head r_1 at the cost of 4.

Given C_i^j for $1 \leq j \leq |R_i|$, the cost of a Shuffle transfer from m senders to a receiver group $R_i = \{r_1, r_2, ..., r_g\}$ is given by

$$C_i = \min\{C_i^1, C_i^2, ..., C_i^{|R_i|}\}. \tag{8.1}$$

As a result, the best entry point for the receiver group R_i can be found simultaneously. It is not necessary that the best entry point for the receiver group R_i is the group head r_1. Accordingly, a Shuffle subgraph from m senders to all receivers in R_i is established. Such a method is called the **SRS-based Shuffle subgraph building method**.

We use an example to reveal the benefit of our SRS-based building method. Consider a Shuffle transfer from all of senders $\{v_2, v_5, v_9, v_{10}, v_{11}, v_{14}\}$ to a receiver group $\{v_0, v_3, v_8\}$ with the group head v_0. Figures 8.1c and 8.2a, b illustrate the IRS-based aggregation trees rooted at v_0, v_3, and v_8, respectively. If one selects the group head v_0 as the entry point for the receiver group, the resultant Shuffle subgraph is of cost 46. When the entry point is v_3 or v_8, the resultant Shuffle subgraph is of cost 48 or 42. Thus, the best entry point for the receiver group $\{v_0, v_3, v_8\}$ should be v_8 not the group head v_0.

In such a way, the best entry point for each of the α receiver groups and the associated Shuffle subgraph can be produced. The combination of such α Shuffle subgraphs generates the final Shuffle subgraph from m senders to n receivers. Therefore, the cost of the resultant Shuffle subgraph, denoted as C, is given by $C = \sum_{I \geq 1}^{\alpha} C_i$.

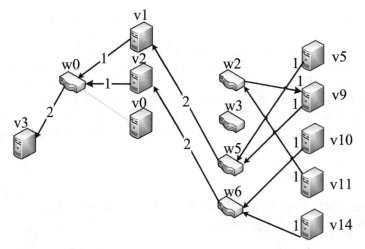

(a) A tree of cost 14 with 11 active links.

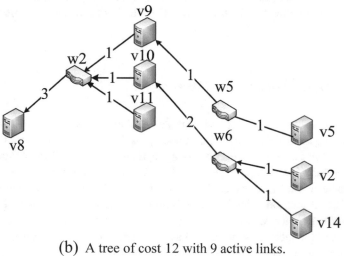

(b) A tree of cost 12 with 9 active links.

Fig. 8.2 Examples of aggregation trees for two incast transfers, with the same sender set and different receivers

8.2.4 The Fault-Tolerance of Shuffle Aggregation Subgraph

In data centers, each Shuffle aggregation subgraph can be impacted by failures on links, servers, or switches. Thus, the fault-tolerance is essential for Shuffle aggregation subgraph. Note that, the fault-tolerant protocol in BCube provides multiple disjoint paths between any server pair. A server at stage j may fail to send data to its parent server due to failures of links or switches. In this case, this server will send data to its parent server which is at stage $j - 1$ along another disjoint path bypassing the failure path.

If a parent server fails, servers cannot send data to it even along alternative paths. In this case, data can be rerouted to next aggregating servers or receivers along other paths if the parent server is not an aggregating server. However, if this parent server is an aggregating server, the intermediate data may have been received and stored already by this failed node. Thus, for senders transmitting data through this node, they need to retransmit data along other disjoint paths to the next aggregating server or receiver.

8.3 Scalable Forwarding Schemes for Performing In-Network Aggregation

We start with a general forwarding scheme to perform the in-network aggregation based on a Shuffle subgraph. Accordingly, we propose two more scalable and practical forwarding schemes based on in-switch Bloom filters and in-packet Bloom filters, respectively.

8.3.1 General Forwarding Scheme

As MapReduce-like systems grow in popularity, there is a strong need to share data centers among users that submit MapReduce jobs. Let β be the number of such jobs running in a data center concurrently. For a job, there exists a JobTracker that is aware of the locations of m Map tasks and n Reduce tasks of a related Shuffle transfer. The JobTracker, as a Shuffle manager, then calculates an efficient Shuffle subgraph using our SRS-based method. The Shuffle subgraph contains n aggregation trees since such a Shuffle transfer can be normalized as n Incast transfers. The concatenation of the job *id* and the receiver *id* is defined as the identifier of an Incast transfer and its aggregation tree. In such a way, we can differentiate all Incast transfers in a job or across jobs in a data center.

Conceptually, given an aggregation tree all involved servers and switches can conduct the in-network aggregation for the Incast transfer via the following operations. Each sender (a leaf vertex) greedily delivers a data flow along the tree if its output for the receiver has generated and it is not an aggregating server. If a data flow meets an aggregating server, all packets will be cached. Upon the data from all its child servers and itself have been received, an aggregating server performs the in-network aggregation on multiply flows as follows. It groups all key-value pairs in such flows according to their keys and applies the aggregate function to each group. Such a process finally replaces all participant flows with a new one that is continuously forwarded along the aggregation tree. Obviously, the size of new flow generated by aggregating server is less than the cumulative size of all input flows.

To put such a design into practice, we need to make some efforts as follows. For each of all Incast transfers in a **Shuffle** transfer, the Shuffle manager makes all servers and switches in an aggregation tree are aware of that they join the Incast transfer. Each device in the aggregation tree thus knows its parent device and adds the Incast transfer identifier into the routing entry for the interface connecting to its parent device. Note that the routing entry for each interface is a list of Incast transfer identifiers. When a switch or server receives a flow with an Incast transfer identifier, all routing entries on its interfaces are checked to determine where to forward or not.

Although such a method ensures that all flows in a Shuffle transfer can be successfully routed to destinations, it is insufficient to achieve the inherent gain of in-network aggregation. Actually, each flow of an Incast transfer is just forwarded even it reaches an aggregating server in the related aggregation tree. The root cause is that a server cannot infer from its routing entries that it is an aggregating server and no matter knows its child servers in the aggregation tree. To tackle such an issue, we let each server in an aggregation tree maintain a $(id, value)$ pair that records the Incast transfer identifier and the number of its child servers. Note that a switch is not an aggregation server and thus just forwards all received flows.

In such a way, when a server receives a flow with an Incast transfer identifier, all $(id, value)$ pairs are checked to determine whether to cache the flow for the future aggregation. That is, a flow needs to be cached if the related $value$ in $(id, value)$ pair exceeds one, and if the $value$ equals the number of flows it received from all its child servers and itself, the server aggregates all cached flows for the same Incast transfer as a new flow. All routing entries on its interfaces are then checked to determine which interface the new flow should be sent out. If the response is null, the server is the destination of the new flow. If a flow reaches a non-aggregating server, it just checks all routing entries to identify an interface the flow should be forwarded.

8.3.2 In-Switch Bloom Filter Based Forwarding Scheme

One trend of modern data center designs is to utilize a large number of low-end switches for interconnection for economic and scalability considerations. The space of TCAM (ternary content addressable memory) in such kind of switches is relatively narrow and thus is quite challenging to support massive Incast transfers by keeping the Incast routing entries. To address such a challenge, we propose two types of Bloom filter based Incast forwarding schemes, namely, in-switch Bloom filter and in-packet Bloom filter.

A Bloom filter is a data structure used to represent set information and determine set membership. It consists of a vector of m bits, initially all set to 0. It encodes each item in a set X by mapping it to h random bits in the bit vector uniformly via h hash functions. To judge whether an element x belongs to X, one just needs to check whether all hashed bits of x in the Bloom filter are set to 1. If not, x is definitely not a member of X. Otherwise, we infer that x is a member of X. A Bloom filter may yield a false positive due to hash collisions, for which it suggests that an element x is in X even though it is not. The false positive probability is $f_p \approx (1 - e^{-h \times n_0/m})^h$. From Literature [5], f_p is minimized as $0.6185^{m/n_0}$ when $h = (m/n_0) \ln 2$.

For the in-switch Bloom filter, each interface on a server or a switch maintains a Bloom filter that encodes the identifiers of all Incast transfers on the interface. When the switch or server receives a flow with an Incast transfer identifier, all the Bloom filters on its interfaces are checked to determine which interface the flow should be sent out. If the response is null at a server, it is the destination of the packet. Consider that a server is a potential aggregating server in each Incast transfer. When a server receives a flow with an Incast identifier, it first checks all $(id, value)$ pairs to determine whether to cache the flow for the future aggregation. Only if the related $value$ is 1 or a new flow is generated by aggregating cached flows, the server checks all Bloom filters on its interfaces for identifying the correct forwarding direction.

The introduction of Bloom filter can considerably compress the forwarding table at each server and switch. Moreover, checking a Bloom filter on each interface only incurs a constant delay, only h hash queries, irrespective of the number of items represented by the Bloom filter. In contrast, checking the routing entry for each interface usually incurs $O(\log \gamma)$ delay in the general forwarding scheme, where γ denotes the number of Incast transfers on the interface. Thus, Bloom filters can significantly reduce the delay of making a forwarding decision on each server and switch. Such two benefits help realize the scalable Incast and Shuffle forwarding in data center networks.

8.3.3   In-Packet Bloom Filter Based Forwarding Scheme

Both the general and the in-switch Bloom filter based forwarding schemes suffer non-trivial management overhead due to the dynamic behaviors of Shuffle transfers. More precisely, a new scheduled Mapreduce job will bring a Shuffle transfer and an associated Shuffle subgraph. All servers and switches in the Shuffle subgraph thus should update its routing entry or Bloom filter on related interface. Similarly, all servers and switches in a Shuffle subgraph have to update their routing table once the related Mapreduce job is completed. To avoid such constraints, we propose the in-packet Bloom filter based forwarding scheme.

Given a Shuffle transfer, we first calculate a Shuffle subgraph by invoking our SRS-based building method. For each flow in the Shuffle transfer, the basic idea of our new method is to encode the flow path information into a Bloom filter field in the header of each packet of the flow. The routing path of a flow is a sequence of links. For example, a flow path from a sender v_{14} to the receiver v_0 in Fig. 8.1c is a sequence of links, denoted as $v_{14} \rightarrow w_6$, $w_6 \rightarrow v_2$, $v_2 \rightarrow w_0$, and $w_0 \rightarrow v_0$. Such a set of links is then encoded as a Bloom filter via the standard input operation [5, 6]. Such a method eliminates the necessity of routing entries or Bloom filters on the interfaces of each server and switch. To differentiate packets of different Incast transfers, all packets of a flow is associated with the identifier of an Incast transfer the flow joins.

When a switch receives a packet with a Bloom filter field, its entire links are checked to determine along which link the packet should be forwarded. For a server, it should first check all $(id, value)$ pairs to determine whether to cache the flow for the future aggregation. That is, the packet should be forwarded directly if the related $value$ is 1. Once a server has received all $value$ flows of an Incast transfer, e.g., server v_{10} collects $value = 3$ flows in Fig. 8.2b, such flows are aggregated as a new flow. To forward such packets, the server checks the Bloom filter with its entire links as inputs to find a forwarding direction.

Generally, such a forwarding scheme may incur false positive forwarding due to the false positive feature of Bloom filters. When a server in a flow path meets a false positive, it will forward flow packets towards another one-hop neighbors besides its right upstream server. The flow packets propagated to another server due to a false positive will be terminated with high probability since the in-packet Bloom filter just encodes the right flow path. Additionally, the number of such mis-forwarding can be less than one during the entire propagation process of a packet if the parameters of a Bloom filter satisfy certain requirements.

For a Shuffle transfer in BCube(n, k), a flow path is at most $2(k + 1)$ of length, i.e., the diameter of BCube(n, k). Thus, a Bloom filter used by each flow packet encodes $n_0 = 2(k + 1)$ directed links. Let us consider a forwarding of a flow packet from stage i to stage $i - 1$ for $1 \leq i \leq k$ along the Shuffle subgraph. The server at stage i is an i-hop neighbor of the flow destination since their labels differ in i dimensions. When the server at stage i needs to forward a packet of the flow, it just needs to check i links towards its i one-hop neighbors that are also $(i - 1)$-hops neighbors of the flow destination. Among the i links on the server at stage i, $i - 1$ links may cause false positive forwarding at a given probability when checking the Bloom filter field of a flow packet. Although the switch has n links, only one of them connects with the server at stage $i - 1$ in the path encoded by a Bloom filter field in the packet. The servers along with other $n - 2$ links of the switch are still i-hops neighbors of the destination. Such $n - 2$ links are thus ignored when the switch makes a forwarding decision. Consequently, no false positive forwarding appears at a switch.

In summary, given a Shuffle subgraph a packet will meet at most k servers before reaching the destination and may occur false positive forwarding on each of at most $\sum_{i=1}^{k-1} i$ server links at probability f_p. Thus, the number of resultant false positive forwarding due to deliver a packet to its destination is given by

$$f_p \times \sum_{i=1}^{k-1} i = 0.6185^{\frac{m}{2(k+1)}} \times k \times (k - 1)/2, \qquad (8.2)$$

here $n_0 = 2(k + 1)$. If we impose a constraint that the value of Formula (8.2) is less than 1, then

$$m \geq 2(k + 1) \times \log_{0.6185} \frac{2}{k \times (k - 1)}. \qquad (8.3)$$

We can derive the required size of Bloom filter field of each packet from Formula (8.3). We find that the value of k is usually not large for a large-scale data center. Thus, the Bloom filter field of each packet incurs additional traffic overhead.

8.4   Performance Evaluation

8.4.1   The Prototype

Our testbed consists of 61 virtual machines (VM) hosted by 6 servers connected with an Ethernet. Each server equips with two 8-core processors, 24 GB memory and a 1 TB disk. Five of the servers run 10 virtual machines as the Hadoop virtual slave nodes, and the other one runs 10 virtual slave nodes and 1 master node that acts as the Shuffle manager. Each virtual slave node supports two Map tasks and two Reduce tasks. We extend the Hadoop to embrace the in-network aggregation on any Shuffle transfer. We launch the built-in wordcount job of Hadoop 0.21.0 with the combiner, where the job has 60 Map tasks and 60 Reduce tasks, respectively. That is, such a job employs one Map task and Reduce task from each of 60 VMs. We associate each Map task ten input files each with 64MB. A Shuffle transfer from all 60 senders to 60 receivers is thus achieved. The average amount of data from a sender (Map task) to the receiver (Reduce task) is about 1MB after performing the combiner at each sender. Each receiver performs the count function. Our approaches exhibit more benefits for other aggregation functions, e.g., the SUM, MAX, MIN, Top-K and KNN functions.

To deploy the wordcount job in a BCube(6, k) data center for $2 \leq k \leq 8$, all of senders and receivers (the Map and Reduce tasks) of the Shuffle transfer are randomly allocated with a $k + 1$-dimensional BCube labels. We then generate the SRS-based, Incast-based, and Unicast-based Shuffle subgraphs for the Shuffle transfer via corresponding methods. Our testbed is used to emulate a partial BCube(6,k) on which the resultant Shuffle subgraphs can be conceptually deployed. To this end, given a Shuffle subgraph, we omit all of switches and map all of intermediate server vertices to the 60 slave VMs in our testbed, while the master VM is used as the Shuffle manager. That is, we use a software agent to emulate an intermediate server so as to receive, cache, aggregate, and forward packets. Thus, we achieve an overlay implementation of Shuffle transfer. Each path between two neighboring servers in the Shuffle subgraph is mapped to a virtual link between VMs or a physical link across servers in our testbed. It is required that all servers mapped to the same VM cannot contain any neighbor pair in the successive stages in Shuffle subgraph. Thus, in the processing of Shuffle transfer, no local communication will be generated among VM agents.

We compare our SRS-based Shuffle subgraph against the Incast-based Shuffle subgraph, the Steiner-based one, the Unicast-based one, and the existing method in terms of four metrics. They are the resultant network traffic, the number of active links, the number of cache servers, and the input data size at each receiver. The network traffic denotes the sum of network traffic over all edges in the Shuffle subgraph.

The Steiner-based Shuffle subgraph is similar to the Incast-based one but each aggregation tree results from the Steiner-tree algorithm. The Steiner-tree algorithm we choose is the one described in Literature [7], whose benefit is the computation speed. In addition, the existing Shuffle subgraph without in-network aggregation is similar to the aforementioned Unicast-based Shuffle subgraph.

8.4.2 Impact of the Data Center Size

Consider a Shuffle transfer with 60 senders and 60 receivers that are randomly selected in a data center with BCube(6,k) as its network structure. We conduct experiments and collect the performance metrics after completing such a Shuffle transfer along different Shuffle subgraphs. Figure 8.3 shows the changing trends of the performance metrics under different methods and settings of k.

Figure 8.3a indicates that our SRS-based, the Steiner-based, and the Unicast-based methods considerably save the resultant network traffic compared to the existing method, irrespective of the data center size. Table 8.1 shows that the SRS-based method causes less network traffic than the Incast-based method. More precisely, the SRS-based, Incast-based, Steiner-based, and Unicast-based methods save the network traffic by 32.87, 32.69, 28.76, and 17.64% on average compared to the existing method. Such results demonstrate the large gain of in-network aggregation in BCube(6,k) even for a small Shuffle transfer.

Additionally, the SRS-based and Incast-based methods considerably outperform the Unicast-based one due to the following reason. The first two methods aggregate all of involved flows at the level of a single or a group of Incast transfers while the Unicast-based method does it at the level of each individual flow. Actually, the number of aggregating servers in the Incast-based method increases while that in the Unicast-based method decreases along with the increase of k, as shown in Fig. 8.3c. Thus, the Incast-based method always has more opportunities to performing the in-network aggregation than the Unicast-driven method. The SRS-based method improves the Incast-based one by reusing as many Incast aggregation trees as possible and thus utilizes less number of aggregating servers and active links, as shown in Fig. 8.3b, c. Note that the SRS-based method achieves the same effect as the Incast-based method

Table 8.1 The network traffic under different k

Shuffle	$k = 2$	$k = 3$	$k = 4$	$k = 5$	$k = 6$	$k = 7$
SRS-based	8544	12,250	20,102	25,608	30,806	37,754
Incast-based	8730	12,298	20,098	25,606	30,806	37,754

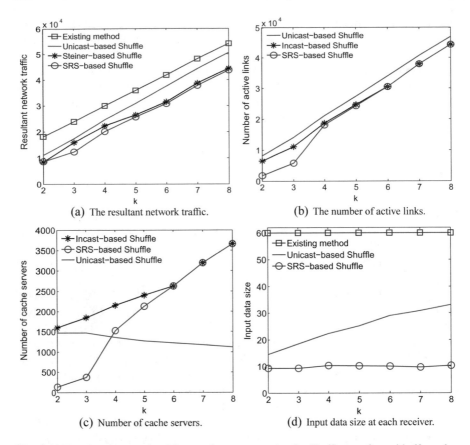

Fig. 8.3 The changing trends of four performance metrics for Shuffle transfers with 60 senders and 60 receivers in BCube(6,k)

when $k \geq 6$. The reason is that the data center is too large such that a small number of receivers, e.g., randomly selected 60 receivers, do not contain any pair of one-hop neighbors. Besides such benefits, the SRS-based method significantly reduces the size of input data at each receiver, as shown in Fig. 8.3d, and thus reduces the delay during the reduce phase of a job.

In summary, the SRS-based and Incast-based methods support a small size Shuffle transfer well with less data center resources and network traffic compared to the Unicast-based and existing methods, irrespective of the size of the data center. Moreover, the SRS-based and Incast-based methods can improve the performance of the Steiner-based method at some extent. The root cause is that the first two methods can efficiently exploit the topological feature of BCube and utilize more aggregating servers than the Steiner-based method although they have similar number of active links.

8.4.3   Impact of the Shuffle Transfer Size

Consider that a MapReduce-like job may sometimes involve several hundreds even thousands Map and Reduce tasks. We will evaluate our methods over Shuffle transfers with varying number of members. It is difficult for our testbed to execute a large-scale wordcount job due to its limited resources. We thus conduct extensively simulations to demonstrate the scaling property of our methods. To achieve a Shuffle transfer with m senders and n receivers, $m = n = 50 \times i$ for $1 \le i \le 30$, we created a synthetic wordcount job. The wordcount job provides the input data of 64 MB for each sender, generated by the built-in RandomTextWriter of Hadoop. The data transmission from each sender to the receiver is controlled to be 1 MB on average. Figure 8.4 shows the changing trends of the performance metrics along with the increase of the number

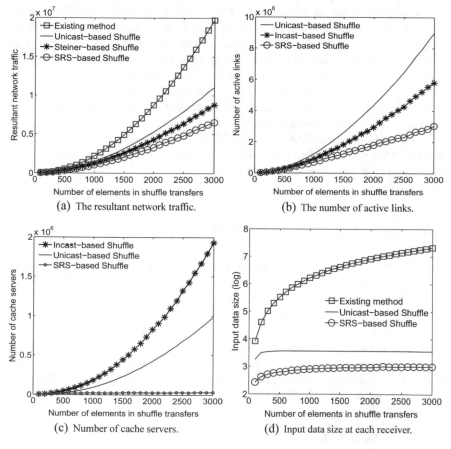

(a) The resultant network traffic.

(b) The number of active links.

(c) Number of cache servers.

(d) Input data size at each receiver.

Fig. 8.4 The changing trends of four performance metrics along with the increase of Shuffle transfer size in BCube(8,5)

of elements, i.e., $m + n$, in Shuffle transfers in BCube(8,5). The number of servers in BCube(8,5) is 262,144 that is large enough for a data center.

Figure 8.4a indicates that the SRS-based, Incast-based, Steiner-based, and Unicast-based Shuffle methods significantly save the network traffic by 55.33, 55.29, 44.89, and 34.46% on average compared to the existing method as $m = n$ ranges from 50 to 1500. Such benefits of the three methods are more notable as the increase of $m = n$ and thus demonstrate the gain of the in-network aggregation in large Shuffle transfers. Moreover, the SRS-based method always utilizes less links (hence less servers and network devices) compared to the Unicast-based and Incast-based methods, as shown in Fig. 8.4b.

On the other hand, the Incast-based method always exploits more aggregating servers than the Unicast-based method as the Shuffle size increases, as shown in Fig. 8.4c. Thus, the Incast-based method always has more opportunity to performing the in-network aggregation than the Unicast-driven method. This is the fundamental reason why the Incast-based method causes less network traffic than the Unicast-based method. The SRS-based method improves the Incast-based one by reusing as many Incast aggregation trees as possible. Consequently, the SRS-based method achieves the largest gain of the in-network aggregation with the least number of aggregating servers and active links compared to the Incast-based method, as shown in Fig. 8.4b, c. Besides the above benefits, the SRS-based method significantly reduces the size of input data at each receiver, as shown in Fig. 8.4d, and thus reduces the delay during the reduce phase of a job.

In summary, the SRS-based and Incast-based approaches can support a Shuffle transfer of any size well at the cost of less data center resources and network traffic compared to the Unicast-based and existing methods.

8.4.4   Impact of the Aggregation Ratio

Recall that we make an assumption that data flows each of which consists of key-value pairs can be aggregated as a new one whose size is the largest size among such flows. That is, the set of keys in each involved data flow is the subset of that in the largest data flow. We further evaluate the SRS-based approach under a more general Shuffle transfer.

Given s data flows towards to the same receiver in a general Shuffle transfer, let f_i denote the size of the ith data flow for $1 \leq i \leq s$. Let δ denote the aggregation ratio among any number of data flows, where $0 \leq \delta \leq 1$. After aggregating such s flows, the size of the new data flow is given by

$$\max\{f_1, f_2, ..., f_s\} + \delta \times \left(\sum_{i=1}^{s} f_i - \max\{f_1, f_2, ..., f_s\} \right).$$

Our analysis in Sect. 8.2 and the above large-scale simulations fall into a special scenario, i.e., $\delta = 0$, where the in-network aggregation on flows achieves the largest gain. On the contrary, $\delta = 1$ is another special scenario where any two of the s data flows do not share any key. In such a case, the in-network aggregation on the s data flows does not bring any gain. Actually, these two extreme scenarios are rare in practice. Here, we evaluate the SRS-based approach in a more general scenario where $0 \le \delta \le 1$.

We measure the resultant network traffic of the SRS-based, Unicast-based, and existing methods under two representative Shuffle transfers in BCube(8,5) as δ ranges from 0 to 1. Figure 8.5 indicates that our SRS-based method always incurs much less network traffic than other two methods. Assume that the value of the random variable δ follows a uniform distribution. In such a case, the SRS-based method saves the resultant network traffic by 28.78% in Fig. 8.5a and 45.05% in Fig. 8.5b compared to the existing method, respectively. Thus, the SRS-based method outperforms other two methods in more general scenarios, irrespective of the size of Shuffle transfer.

We further compare the SRS-based Shuffle with SRS-based Shuffle (no agg.) methods in terms of the completion time of MapReduce jobs on average. As expected, when the aggregation ration approaches to 1 the performance of such two methods is the same, because there is no opportunity to aggregate packets on-path. As aggregation ration decreases, our SRS-based Shuffle method always achieves the lower Shuffle and reduces time due to reduce the number of packets forward and increase the available bandwidth. This is clearly seen in Fig. 8.6.

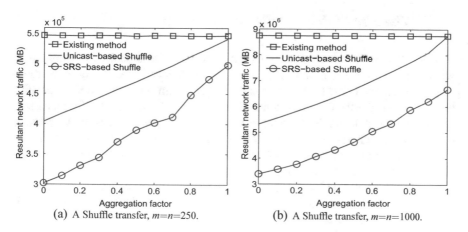

(a) A Shuffle transfer, $m=n=250$.　　(b) A Shuffle transfer, $m=n=1000$.

Fig. 8.5 The changing trends of the network traffic along with the increase of aggregation factor in BCube(8,5)

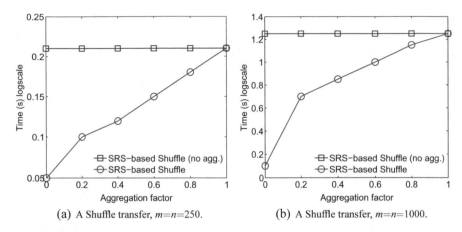

(a) A Shuffle transfer, $m=n=250$.　　　　　(b) A Shuffle transfer, $m=n=1000$.

Fig. 8.6 The changing trends of the delay along with the increase of aggregation factor of Shuffle transfers in BCube(8,5)

Table 8.2 The minimum size of bloom filter field in each packet

	$k=2$	$k=3$	$k=4$	$k=5$	$k=6$	$k=7$
Bits (m)	<19	19	38	58	79	102
BCube(6,k)	216	1296	7776	46,656	279,936	1,679,616

8.4.5 The Size of Bloom Filter in Each Packet

Among the three forwarding schemes for performing the in-network aggregation, we prefer the in-packet Bloom filter based forwarding scheme. The only overhead of such a scheme is the traffic overhead resulting from not only the false positive forwarding but also the Bloom filter field in each packet. We reveal that, for a packet of any Shuffle transfer, no false positive forwarding occurs during the entire forwarding process along with a SRS-based Shuffle subgraph only if Formula (8.3) holds.

For data centers with network structures BCube$(6, k)$s, Table 8.2 shows the minimum size of Bloom filter field in each packet. The traffic overhead due to the Bloom filter field in each packet increases along with the increase of k. The overhead, however, is less than 10 bytes in data centers, with no more than 279,936 servers, that is large enough for a production data center.

References

1. Condie T, Conway N, Alvaro P, et al. MapReduce Online [C]. In Proc. of 7th USENIX NSDI, San Jose, USA, 2010: 313–328.
2. Guo C, Lu G, Li D, et al. BCube: A high performance, server-centric network architecture for modular data centers[J]. ACM SIGCOMM Computer Communication Review, 2009, 39(4): 63–74.
3. Chowdhury M, Zaharia M, Ma J, et al. Managing data transfers in computer clusters with orchestra [J]. ACM SIGCOMM Computer Communication Review, 2011, 41(4): 98–109.
4. Chen Y, Ganapathi A, Griffith R, et al. The case for evaluating mapreduce performance using workload suites [J]. In Proc. of 19th IEEE/ACM MASCOTS, Singapore, 2011: 390–399.
5. Guo D, Wu J, Chen H, et al. The dynamic bloom filters [J]. IEEE Transactions on Knowledge and Data Engineering, 2010, 22(1): 120–133.
6. Broder A, Mitzenmacher M. Network applications of bloom filters: A survey [J]. Internet Mathematics, 2005, 1(4): 485–509.
7. Kou L, Markowsky G, Berman L. A fast algorithm for Steiner trees [J]. Acta Informatica (Historical Archive), 1981, 15(2): 141–145.

Chapter 9
Collaborative Management of Uncertain Incast Transfer

Abstract We have discussed the in-network aggregation problems of Incast and Shuffle transfers in Chaps. 7 and 8 in detail. Note that these problems are only considered after an Incast or Shuffle transfer has been generated; that is, the sender and receiver of each flow are determined. However, there are diverse compute nodes and storage nodes to select for many data center applications, and different selections lead to different senders and receivers of a corresponding Incast transfer. This type of correlated transfers is defined as *uncertain Incast transfer*. Compared with deterministic Incast, uncertain Incast has the opportunity to achieve more in-network aggregation gain. Prior approaches, relying on deterministic incast transfers, remain inapplicable. This chapter makes the first step towards the study of aggregating uncertain Incast transfer. We propose efficient approaches from two aspects, i.e., the initialization of uncertain senders and the Incast tree building. We design two initialization methods to pick the best deterministic senders for an uncertain Incast transfer to form the least number of disjoint sender groups. Thus, flows from each group would be aggregated as one flow on a common one-hop neighbor, irrespective of the location of a picked receiver. Accordingly, we propose two Incast tree-building methods to exploit the benefits of our initialization methods. Experiments show that an uncertain Incast transfer significantly outperforms any related deterministic one in terms of the reduced network traffic and the saved network resources.

9.1 Introduction

The in-network aggregation of correlated transfers aims to push the aggregation computation into the network. Through the collaborative design of transmission paths for a group of correlated transfers, the in-network cache and aggregation can be implemented at some rendezvous devices. As discussed in Chaps. 7 and 8, aggregating correlated transfers can significantly reduce the resultant traffic amount in the transmission stage and save the network resources. In Chap. 7, the in-network aggregation of an Incase transfer is discussed in the situation, where the endpoints of each flow in an Incast transfer are fixed in advance. We call this case the *deterministic Incast transfer*. Only in this way, all flows of an Incast transfer can be scheduled along an

© Springer Nature Singapore Pte Ltd. 2022
D. Guo, *Data Center Networking*,
https://doi.org/10.1007/978-981-16-9368-7_9

optimal Incast tree to exploit the gain of the in-network aggregation via caching and aggregating flows during the transmission process.

Unfortunately, in many cases, the characteristics of an Incast transfer are unknown prior. There are diverse compute nodes and storage nodes for many data center applications to select, and different selections lead to different senders and receivers of the corresponding Incast transfer. Actually, for many Incast transfers, it is not necessary that the endpoints of their flows have to be at specific locations as long as certain constraints are satisfied. An illustrative scenario is that GFS-like file systems will write each block of large pieces of data into three machines across two racks. Thus, a map task would be scheduled to one of three machines hosting its input data to achieve the data locality. This fact brings the sender flexibility for an Incast transfer. More precisely, for an Incast transfer with m flows, the number of different combinations of possible sender endpoints is 3^m. Moreover, a reduced task may be scheduled to any machine in a data center as long as it has sufficient task slots. This fact incurs the receiver's flexibility for an Incast transfer. Such a problem is defined as the *uncertain Incast transfer*, a more general setting of Incast transfer. The uncertain sender of each flow refers to an endpoint set, not a fixed endpoint. So does the uncertain receiver. This brings new opportunities to lower down the network traffic per Incast transfer.

In this chapter, our objective is to minimize the amount of the network traffic for completing an uncertain Incast transfer by applying the in-network aggregation. In reality, an uncertain Incast transfer will be embodied as any possible deterministic Incast transfer via initializing each flow's uncertain sender and receiver as deterministic ones. The resultant deterministic Incast transfers have diverse gains of in-network aggregation. A simple way is to apply the in-network aggregation approach in Chap. 7 to each potential deterministic Incast transfer and select the aggregation tree with the most aggregation gain. Such a way suffers the complexity of solving a large number of NP-hard problems. It has been shown that discovering a minimal aggregation tree for a single deterministic Incast transfer is NP-hard. Prior approaches, relying on deterministic Incast transfers, remain inapplicable to uncertain Incast transfers.

We present a novel aggregation approach for uncertain Incast transfers from two stages to reduce resultant network traffic based on the above analysis. In the first stage, we model the initialization of uncertain senders as the minimal sender group (MSG) problem. That is, those picked best senders for all involved flows form the least number of disjoint groups, where all senders in each group are mutual one-hop neighbors. Thus, flows from the same group would be aggregated as one by a common one-hop neighbor, irrespective of the picked receiver. This would reduce the caused network traffic at the earliest time. In this chapter, we propose two efficient methods, the SD-based and MD-based MSG methods, to achieve such benefits.

In the second stage, given the set of picked best senders and a random receiver, we build an aggregation Incast tree with the least cost. However, approaches proposed in Chap. 7 fail to exploit the benefits of our MSG initialization methods and achieve less gain of in-network aggregation. In this chapter, we design two approximate methods for uncertain Incast transfers, the *interstage-based* and *intrastage-based* methods.

They can build an efficient Incast aggregation tree by exploiting the benefits of the preferred MD-based initialization method.

We evaluate the proposed approaches with prototype experiments and large-scale simulations. The results demonstrate that these approaches can achieve the benefits of in-network aggregation for any uncertain Incast transfer, significantly reducing the caused network traffic and saving network resources. Moreover, the intrastage-based and interstage-based Incast trees for an uncertain Incast transfer can save the caused network traffic up to 33.85 and 27% on average, against the prior IRS-based Incast tree. We can also see from the results that the intrastage-based method achieves a better aggregation effect than interstage-based method. Furthermore, network traffic can be saved if the members of an Incast transfer in the data centers are selected in a managed way instead of a random way.

9.2 Overview of Aggregating Uncertain Incast Transfer

We start with the definition and explanation of uncertain Incast transfer, then measure the effect of aggregating an Incast transfer. Finally, we discuss the problem of aggregating an uncertain Incast transfer.

9.2.1 Problem Statement of Uncertain Incast Transfer

For a data center, we model it as a graph $G = (V, E)$ with a vertex set V and an edge set E. A vertex of the graph corresponds to a switch or a datacenter server. An edge (u, v) denotes a link, through which u connects with v where $v, u \in V$.

As mentioned in Chaps. 7 and 8, many data-intensive frameworks, such as MapReduce, Dryad, Pregel, and Spark, incur serious Incast transfers across successive stages in data centers. Moreover, such Incast transfers usually consume a large number of network resources and make bandwidth resource the bottleneck of data centers. Prior proposals [1, 2] for accelerating the completion of an Incast transfer, however, assumes prior knowledge of transfer characteristics. That is, the endpoints of each flow in an Incast transfer are fixed in advance. This is called the deterministic Incast transfer.

Actually, for many Incast transfers, it is not necessary that the endpoints of their flows have to be in specific locations as long as certain constraints are satisfied. For example, a Map task would be scheduled to one of three machines hosting its input data to achieve the data locality. This brings the sender flexibility for an Incast transfer. To ease the presentation, the number of the possible endpoint of the source of each flow is set to three. Thus, for an Incast transfer with m flows, the number of different combinations of possible sender endpoints is 3^m. Moreover, each Reduce task will be scheduled in the entire data center as long as certain constraints can be satisfied. This incurs the receiver's flexibility for an Incast transfer. Based on the above analysis, we can see that the sender or the receiver of an Incast transfer usually represents a server group rather than a specific server. This general problem is formalized as an *uncertain Incast transfer*, as defined in Definition 9.1.

Definition 9.1 An uncertain Incast transfer consists of m flows from a set of senders $\{s_1, s_2, \ldots, s_m\}$ to the receiver R. The uncertain sender s_i of each flow refers to an endpoint set S_i, each member in which can be selected as the sender of ith flow. Moreover, the common receiver of these m flows is also uncertain, which can be selected from a set of servers.

A *weak* uncertain Incast transfer is a special case of uncertain Incast transfer with a deterministic receiver or deterministic senders. When both the senders and the receiver are determined, we achieve a deterministic Incast transfer, *abbreviated as Incast transfer* in the remainder of this chapter. As aforementioned, all highly correlated flows of an Incast transfer should be aggregated during the transmission process. Before discussing the in-network aggregation of an uncertain Incast transfer, we start with the Incast transfer as an example.

9.2.2 Aggregating a Deterministic Incast Transfer

For any Incast transfer, it is feasible to significantly reduce the amount of caused traffic by performing the in-network aggregation, thereby greatly saving the network resources. More precisely, we need to construct an Incast tree from the graph $G = (V, E)$ by combining those Unicast paths from all senders to the same receiver. The flow from each sender thus can be delivered to the receiver along the formed tree. Consider that there exist multiple equal-cost routing paths between any pair of nodes in densely connected data center networks. That is, a flow between any pair of sender and receiver can choose any route from multiple ones, for example, α. In this way, the selected routing paths for such flows generate at most α^m potential Incast trees, given an Incast transfer with m senders. To select a desired Incast tree, with the minimal network traffic due to utilize the in-network aggregation, a cost metric should be defined for measuring each Incast aggregation tree.

As mentioned in Chap. 2, existing network topologies for data centers can be roughly divided into two categories, switch-centric and server-centric. Given an Incast transfer, they differ in selecting and using aggregating nodes in related Incast tree. Hence, the cost metric of an aggregation tree should be calculated in a reliable way respectively.

For server-centric network topologies, such as BCube, each vertex in an Incast tree represents either a commodity server or switch. Commodity servers support the in-network cache and advanced packet processing and hence deliver a programmable data plane. The used commodity switches, however, cannot naturally support a programmable data plane. Thus, a vertex is an aggregating vertex only if it represents a server and at least two flows converge at it. It then aggregates its incoming flows and forwards a generated single flow instead of multiple individual flows along the tree. Usually, we assume that the size of the resultant single flow is the maximum size among all incoming flows of the aggregating vertex, while at a non-aggregating vertex, the size of its outgoing flow is the sum size of its incoming flows.

In server-centric data centers, commodity switches are treated as non-aggregating vertices. However, novel switches, which already have enough resources and abilities for aggregating flows, can indeed be aggregating nodes in switch-centric network topologies, such as Fat-Tree. That is, each non-leaf vertex in an Incast tree represents a switch, which is an aggregating vertex if it meets at least two flows. The leaf vertexes represent commodity servers and are not aggregating nodes.

After clarifying the usage of aggregating nodes in an Incast aggregation tree under both types of network topologies, we present a general cost metric for any Incast aggregation. Specifically, the cost of an Incast tree means the amount of overall traffic for completing the Incast transfer.

Without loss of generality, we assume that the traffic from each sender is 1MB. Thus, the size of the outgoing link is one at an aggregating vertex and equals the number of incoming flows at a non-aggregating vertex. If an Incast tree does not adopt the in-network aggregation, all of the vertices are non-aggregating ones. The gain of an Incast aggregation tree is the cost difference between the trees without and with the in-network aggregation.

9.2.3 Aggregating an Uncertain Incast Transfer

In this chapter, our objective is to minimize the amount of the network traffic for completing an uncertain Incast transfer by applying the in-network aggregation. Due to the flexibility of both senders and the receiver, an uncertain Incast transfer can be treated as a group of Incast transfers. For an Incast transfer, the minimal aggregation tree problem is to find a connected subgraph in $G = (V, E)$ that spans all Incast members with the minimal cost. To discover a minimal aggregation tree for an Incast transfer in BCube, a densely connected data center network, is NP-hard. Accordingly, it is NP-hard to identify a minimal aggregation tree for an uncertain Incast transfer even the receiver is determined.

Without loss of generality, we use BCube as an example to show the aggregation of uncertain Incast transfer. BCube(n, k) is an emulation of a $k + 1$ dimensional n-ary generalized hypercube. In BCube(n, k), two servers, labeled as $x_k x_{k-1} \ldots x_1 x_0$ and $y_k y_{k-1} \ldots y_1 y_0$ are mutual one-hop neighbors in dimension j if their labels only differ in dimension j. Such servers connect to a j level switch with a label $y_k \cdots y_{j+1} y_{j-1} \cdots y_1 y_0$ in BCube(n, k). Therefore, a server and its $n - 1$ 1-hop neighbors in each dimension are connected indirectly via a switch. Additionally, two servers are j-hop neighbors if their labels differ in number of j dimensions. As shown in Fig. 9.1, servers v_0 and v_{15}, with labels 00 and 33, are 2-hop neighbors since their labels differ in 2 dimensions.

To realize the online building of aggregation tree for an uncertain Incast transfer with m flows, we aim to design approximate methods to exploit the topological feature

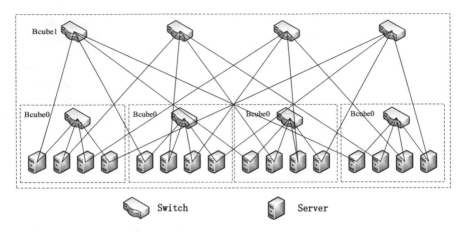

Fig. 9.1 An illustrative example of BCube(4,1)

of datacenter networks in the next section. Such methods involve two successive challenging issues. First, the initialization of uncertain senders aims to pick the best deterministic senders from 3^m candidates for an uncertain Incast transfer if each flow has 3 potential senders. Second, we need to derive the preferred Incast tree for achieving the outstanding aggregation gain, irrespective of the location of the receiver.

In this chapter, BCube is selected as a carrier for aggregating uncertain Incast transfer. Note that the methods proposed in this chapter are still applicable to other server-centric network topologies. Moreover, the approaches proposed in this chapter can be directly applied to FBFLY [3], and HyperX [4], two switch-centric network topologies, if they utilize those novel switches each with a programmable data plane. The reason is that the topology behind BCube is an emulation of the generalized hypercube, while the topologies of FBFLY and HyperX are just the generalized hypercube at the level of switch.

9.3 Aggregation Tree Building Method for Uncertain Incast Transfer

For an uncertain Incast transfer, we start with analyzing the diversity of its aggregation gain and design two initialization methods for uncertain senders. We then design two dedicated Incast tree-building methods to achieve outstanding aggregation gain. Note that the aggregating tree built by these two methods can achieve good gain, irrespective of the initialization methods of the receiver.

9.3.1 The Diversity of In-Network Aggregation Gain

In reality, an uncertain Incast transfer with m flows will be embodied as any potential Incast transfer instance via selecting one from at most 3^m sender sets and assigning a receiver. However, all instances of an uncertain Incast transfer possess different gains of in-network flow aggregation and exhibit diverse tree costs. We explain this with an example shown in Fig. 9.2.

For an uncertain Incast transfer in BCube(4,1), $\{s_1, s_2, s_3, s_4\}$ are four uncertain senders, while r_0 is the common receiver. Assume that $s_1 \in \{v_2, v_5, v_6\}$, $s_2 \in \{v_9, v_{10}, v_{15}\}$, $s_3 \in \{v_1, v_{10}, v_{11}\}$, and $s_4 \in \{v_7, v_{13}, v_{14}\}$. When the sender set is $\{s_1 = v_6, s_2 = v_9, s_3 = v_{10}, s_4 = v_7\}$, the resultant Incast tree is of cost 14 with 11 active links. Thus, all of senders form three groups, $\{v_9\}$, $\{v_6, v_{10}\}$, $\{v_7\}$, as shown in Fig. 9.2a. If the sender set is $\{s_1 = v_5, s_2 = v_9, s_3 = v_{10}, s_4 = v_{14}\}$, the resultant Incast tree is of cost 12 with 9 active links. In this case, those senders incur two groups, $\{v_5, v_9\}$, $\{v_{10}, v_{14}\}$, as shown in Fig. 9.2b. However, a more efficient Incast transfer is of cost 8 with 6 active links if we construct in the way shown in Fig. 9.2c, where $\{s_1 = v_5, s_2 = v_9, s_3 = v_1, s_4 = v_{13}\}$. In this case, four senders form just one group, $\{v_1, v_5, v_9, v_{13}\}$ since they are one-hop neighbors.

As shown in the example above, the three uncertain sender sets cause diverse settings of sender groups. We can see that flows from senders in each group are aggregated as just one flow after one-hop transmission. In this way, the sender set with a smaller number of sender groups will cause less amount of network traffic in the next round of forwarding towards the receiver. To minimize the amount of resultant network traffic, we prefer to find the sender set, which can be partitioned into the least number of sender groups. This design can push the traditional aggregation operation into the transmission process of flows as early as possible; hence, it can significantly save the network resource. This motivates us to study the minimal sender group problem of any uncertain Incast transfer.

9.3.2 Initializing Senders for an Uncertain Incast Transfer

A simple initialization method is to just randomly pick one from the endpoint set of each uncertain sender. As shown in Fig. 9.2, this method cannot ensure to pick the best set of senders for achieving the largest gain of in-network flow aggregation. For this reason, we introduce the MSG problem to replace such a random non-MSG initialization method.

Definition 9.2 For an uncertain Incast transfer with m flows, the minimal sender group (MSG) problem means picking a deterministic one from a set of senders for each flow, and the picked senders form the least number partitioned groups.

The partition method dominates the selection of the sender set. To exploit the benefits of in-network aggregation, we expect that flows originating from all senders

Fig. 9.2 For an uncertain
Incast transfer in
BCube(4,1), we show three
potential aggregating trees

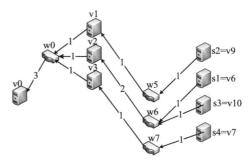

(a) An Incast tree of cost 14 with 11 links.

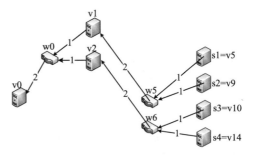

(b) An Incast tree of cost 12 with 9 links.

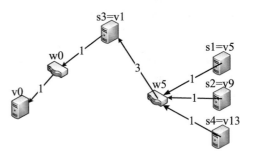

(c) An Incast tree of cost 8 with 6 links.

in each partitioned group reach a common aggregating node after at most one-hop
transmission. Thus, the caused network traffic is significantly reduced. We describe
this design demand in Definition 9.3.

Definition 9.3 For any node-set in BCube(n, k), it can be partitioned if the following
two constraints are satisfied: (1) The intersection among groups is empty; that is,
there is no common node among groups. (2) Any node pair in a group are one-hop
neighbors.

For a sender group in BCube(n, k), we define a partition symbol $e_j \in \{0, 1, \ldots, k\}$
for $0 \le j \le k$. For each e_j in such $k + 1$ partition symbols, all of the senders can

be partitioned into groups such that any two members in the same group are one-hop neighbors in e_j dimension. That is, for each group, the labels of its members only differ in dimension e_j. Clearly, in each resultant group, all members share a common one-hop neighbor, on which their flows are aggregated. This partitioning method meets the constraints in Definition 9.3. Thus, given an Incast transfer, we can partition all of its senders based on partition symbols and compare the partition result of different partition symbols. The partition symbol causing the least number of groups are selected as the optimal partition symbol. This method can be simplified as follows. For any e_j, we extract the label of each sender except the dimension e_j. The resultant label denotes a group covering such a sender. All senders can be partitioned into corresponding groups after they are processed based on the above steps. The computation overhead under e_j is proportional to the number of senders, m. The time complexity is $O(m \times k)$ due to selecting the best partition symbol from $k + 1$.

However, the above efficient partition method is inapplicable to tackle the MSG problem of an uncertain Incast transfer. The reason is that each uncertain sender can be anyone in a given sender set. A straightforward way is to embody an uncertain Incast transfer as 3^m possible Incast transfers and apply the above partition method to each Incast transfer. Accordingly, we select an Incast transfer from all possible instances, which results in the least number of groups. The best solution, however, cannot be derived in polynomial time due to its time complexity $O(m \times k \times 3^m)$. Thus, we design dedicated methods for the MSG problem of an uncertain Incast transfer as follows.

1. **A greedy MSG method based on one dimension**

For an uncertain Incast transfer, the basic idea is to partition the sender set of each flow, denoted as $S_i (1 \leq i \leq m)$, using each partition symbol e_j for $0 \leq j \leq k$ individually. The **Partition** (S, e_j) function in Algorithm 9.1 presents the implementation details. Accordingly, we can find the partition symbol, among $k+1$ ones, which incurs the least number of groups after $k + 1$ rounds of **Partition**(S, e_j) in Algorithm 9.1. Without loss of generality, we assume that such a partition symbol is e_0. The partitioned groups resulting from the symbol e_0, however, are unreasonable for the minimal sender group problem. The root cause is that such partitioned groups contain the whole members of each S_i for $1 \leq i \leq m$, i.e., all possible senders of each flow. For this reason, we design a **Cleanup()** function to refine the partitioned groups such that just one member exists in each S_i.

We first sort the partitioned sender groups according to their cardinalities in descending order and then process each sender group orderly. For each member $node_0$ in the first sender group with the largest cardinality, we verify whether S_i contains $node_0$ for $1 \leq i \leq m$. If it is true, the set S_i just keeps $node_0$ while deleting all other members. This operation makes the uncertain S_i become a deterministic one. Moreover, those deleted members from S_i should also be removed from other corresponding partitioned groups. If none of S_i contains $node_0$, $node_0$ should be removed from that group. For the second-largest group, we perform the same operations afore-mentioned. Such a recursive method finally results in a refined set of sender groups,

---

Algorithm 9.1 The SD-based MSG method

Require: Given an uncertain Incast transfer with m flows in BCube(n,k), where the sender set of each flow is $S_i (1 \leq i \leq m)$

1: Let S denote the union of all members in the sets $S_1, \cdots, S_m$.
2: Let $Lgroups$ be a set of partitioned groups of S, which is empty.
3: **for** $i = 0$ to k **do**
4:     $e_i \leftarrow i$ ▷ i denotes a partition symbol
5:     Tgroups←**Partition**(S, e_i)
6:     **if** $Lgroups$ is empty or $|Tgroups| < |Lgroups|$ **then**
7:         $Lgroups \leftarrow Tgroups$
8: Return the output of **Cleanup**($Lgroups$).

Partition(S, e_j)

1: **while** S contains a sender with label $x_k \cdots x_1 x_0$ **do**
2:     Allocate this sender to a group labeled as $x_k \cdots x_{i+1} x_{i-1} \cdots x_0$.
3:     Remove this sender from the set S.
4: Return the resultant groups.

Cleanup($Lgroups$)

1: Sort $Lgroups$ in the descending order of the set cardinality.
2: **for** $i = 0$ to $|Lgroups|$ **do**
3:     Consider each $node_0$ of the ith group in $Lgroups$. For each $S_j (1 \leq j \leq m)$ containing $node_0$, S_j just keeps $node_0$ while deleting all of other members. All of deleted members from S_j should be removed from other groups, so as to keep the consistency between S and $Lgroups$.

---

which satisfy the requirement of Definition 9.3. That is, each S_i for $1 \leq i \leq m$ just contains one member; hence, the source of each flow becomes deterministic from uncertain. Next, we analyze the time complexity of Algorithm 9.1.

Theorem 9.1 *The time complexity of the SD-based MSG initialization method is* $O(m^2 + m \times k)$.

Proof Algorithm 9.1 consists of two stages, the partition and cleanup. In the first stage, the time complexity of **Partition**(S, e_i) is $O(m)$ since it processes at most $3 \times m$ senders. The entire time complexity is $O(m \times k)$ due to identify the partitioned result with the least number of groups under $k + 1$ partition symbols. In the second stage, the time complexity due to sort groups is $O(m^2)$ since the number of partitioned groups is at most $3 \times m$ in the best case. The time complexity of the cleanup operation is $O(m^2)$, due to compare any pair of partitioned groups. Thus, the entire complexity is $O(m^2 + m \times k)$; hence, Theorem 9.1 is proved.   □

2. A greedy MSG method based on multi-dimensions

Although the SD-based partition method is more efficient than the randomly initializing method, the following topological properties of data centers have not been utilized to enhance its efficiency and performance.

First, each node, $node_0$, may appear at more than one S_i. That is, the intersection of some pairs of S_i are not empty for $1 \leq i \leq m$. This observation is true since a server may host the required data of multiple tasks of a job. If all of such tasks are scheduled to $node_0$, the resulting flows from such tasks can be aggregated locally. This will considerably increase the gain of in-network flow aggregation and save the network resource. Accordingly, such nodes should have priority to be selected by all involved S_i during the partition process in Algorithm 9.1.

Second, all members in S_i for $1 \leq i \leq m$ can be partitioned into the least number of groups using e_0 in the first stage. However, the partitioned result is not necessary to keep the least number of groups after being refined by the cleanup function. As a result, the output of Algorithm 9.1 may hold the least number of groups under other partition symbols. Moreover, some isolated groups, each with only one member under a partition symbol, still have the opportunity to be one-hop neighbors under another partition symbol. For example, two isolated groups $\{v_0\}$ and $\{v_{12}\}$ under e_1 actually can be one-hop neighbors under e_0 in BCube(4,1); hence, they can form a group $\{v_0, v_{12}\}$.

To fully exploit such benefits, we further design Algorithm 9.2, which consists of three major stages, pre-processing, the partition, and the cleanup.

The basic idea of the first stage is to exploit the above first kind of property. Given all of the uncertain sets $S_1, \ldots, S_m$, we first need to derive out the frequency of each unique element. This can be simply realized by traversing all of the elements in each uncertain set at the cost of $O(m)$ time complexity. Note that the sum of cardinalities for $S_1, \ldots, S_m$ is $3 \times m$. For any unique element $node_0$ with non-one frequency, those uncertain sets with it should just keep $node_0$ while removing other members. This can be completed at the cost of $O(m^2)$ time complexity, since each unique element checks at most $m - 1$ uncertain sets, and the number of $node_0$ with non-one frequency is at most $3 \times m$.

To exploit the second kind of property, Algorithm 9.2 needs to carefully re-design the partition and cleanup operations. In the partition stage, the union of $S_1, \ldots, S_m$ is partitioned using each symbol e_i independently, for $0 \leq i \leq k$. Let $Lgroups$ record the combination of partitioned results under all partition symbols, with the cardinality of at most $3k \times m$. The cause is that the number of partitioned groups is at most $3 \times m$ in the worst case for each partition symbol. Accordingly, we can derive that the time complexity of the second stage is $O(k \times m)$.

Algorithms 9.2 and 9.1 differ in the cleanup function with the changed partition stages. In the beginning, all groups in $Lgroups$ are sorted according to their cardinalities in descending order. The time complexity of this process is $O(k^2 \times m^2)$ since $3k \times m$ groups compare their cardinalities with each other in the worst case.

For each $node_0$ in the first sender group with the largest cardinality, it also appears at other k groups due to the partition from $k + 1$ dimensions. Therefore, $node_0$ should be removed from other groups with it. Additionally, we verify whether S_j contains $node_0$ for $1 \leq j \leq m$. If it is true, the set S_j just keeps $node_0$ while deleting all other members. This operation makes the uncertain S_j become a deterministic one. At the same time, those deleted members from S_j have to be removed from those partitioned groups with them. If none of S_i contains $node_0$, $node_0$ should be removed from that group. For other available groups, we recursively perform the same operations in order. Such a recursive method finally results in as few sender groups as possible, which satisfy the requirement of Definition 9.3. We measure the time complexity of the MD-based partition method in Theorem 9.2.

Algorithm 9.2 The MD-based greedy MSG method

Require: An uncertain Incast transfer with m flows in BCube(n,k). The sender set of each flow is
   $S_i (1 \leq i \leq m)$
1: Process the input via calling the **Pre-process** function.
2: Let S denote the union of all members in the sets $S_1, \cdots, S_m$.
3: Let $Lgroups$ record the partition results of S.
4: **for** $i = 0$ to k **do**
5:    $e_i \leftarrow i$ ▷ denotes a partition symbol
6:    Add the output of **Partition**(S, e_i) into $Lgroups$.
7: **Cleanup**$(Lgroups)$
8: Return $Lgroups$

Pre-process$(S_1, \cdots, S_m)$
1: **for** $i = 0$ to m **do**
2:    For any member $node_0$ of S_i, if it also appears in other S_j for $i \neq j$, S_i and S_j keep $node_0$
   while remove all other members.

Cleanup$(Lgroups)$
1: Sort $Lgroups$ in the descending order of the set cardinality.
2: **for** $i = 0$ to $|Lgroups|$ **do**
3:    For any $node_0$ of the ith group in $Lgroups$, all of other groups need to remove it if possible.
4:    For each S_j containing $node_0$ for $1 \leq j \leq m$, S_j just keeps $node_0$ while deleting all of other
   members.
5:    All of deleted members from S_j should be removed from other groups, so as to keep the
   consistancy between S and $Lgroups$.

Theorem 9.2 *The time complexity of the MD-based MSG method is $O(k^2 \times m^2)$.*

Proof Algorithm 9.2 consists of three major stages. For the pre-process stage, the
time complexity is $O(m^2 + m)$. The cause is that each of at most $3 \times m$ unique
element checks at most $m - 1$ uncertain sets. The time complexity of the partition
stage is $O(k \times m)$, due to call $k + 1$ rounds of **Partition**(S, e_i) function, whose time
complexity is $O(m)$. We then calculate the time complexity of the cleanup stage. To
be specific, the sorting operation costs of $O(k^2 \times m^2)$ time complexity. The operation
in line 3 of removing at most $3m$ $node_0$s from potential $3k \times m$ groups is of time
complexity $O(k \times m^2)$. The operation in line 4 of updating related uncertain sets S_i
is of time complexity $O(m^2)$, due to compare at most $3m$ $node_0$ with m uncertain
sets. The final operation in line 5 is of time complexity $O(k \times m^2)$ due to compare
$node_0$, at most number of $3m$, with at most $3k \times m$ potential groups. Thus, the entire
complexity of the third stage is $O(k^2 \times m^2)$. In summary, the entire time of complexity
of three stages is $O(k^2 \times m^2)$. Theorem 9.2 is proved.   □

9.3.3 Aggregation Tree Building Method for Uncertain Incast

The above initialization methods select only one sender for each flow. Such senders
have been organized as a series of groups, $G_1, G_2, \ldots, G_\beta$. In this way, an uncertain

Incast transfer has been transferred as a weak, uncertain Incast transfer with an uncertain receiver and a set of deterministic senders.

After selecting a receiver randomly, we can exploit the in-network aggregation gain of the resultant deterministic Incast transfer. Assuming that the receiver of the Incast transfer is R, and the sender set is $S = \{s_1, s_2, \ldots, s_\alpha\}$. According to Definition 9.3, the partition result of these senders are $G_1, G_2, \ldots, G_\beta$. Note that α is usually less than m since some servers may be senders for more than one flow. For this reason, we use a variable c_i to denote the multiplicity of each server s_i for $1 \leq i \leq \alpha$ in the union of $S_1, \ldots, S_m$.

For such an Incast transfer derived from that uncertain Incast transfer, our objective is to construct a minimal Incast aggregation tree in $G = (V, E)$, that spans all Incast members with minimal cost for completing the Incast transfer. The data flow from each sender thus can be delivered to the receiver R along the formed tree. As mentioned in Chap. 7, building such an Incast tree for any Incast transfer is NP-hard in many data center networks. Prior incast-tree building approaches, relying on deterministic incast transfers, are unaware of the gains of initializing uncertain senders. Accordingly, we design an approximate method to fully exploit the benefits of the initialization result of uncertain senders.

The aggregation Incast tree can be expanded as a directed multistage graph from all of the senders towards the only receiver. The multistage graph is at most $k + 2$ stages. An intrinsic insight for deriving an aggregation incast tree is to first map all of senders into those stages. Each sender s_i for $1 \leq i \leq \alpha$ should appear at the stage j if it is a j-hop neighbor of the receiver. It is clear that the stage 0 only has the receiver. During such a mapping process, the associated groups $G_1, \ldots, G_\beta$ of the sender set S exhibit the following characteristics. For any G_i for $1 \leq i \leq \beta$, the distance from it to the receiver r is defined as j accordingly.

1. If $|G_i| = 1$, the only sender just appears at the stage j if it is a j-hop neighbor of the receiver.
2. If $|G_i| > 1$, all of its senders are mutual one-hop neighbors in a given dimension. At most, one sender is $(j-1)$-hop neighbor while other senders are j-hop neighbors of the receiver. Thus, all senders of G_i appear at the same stage j or across stages j and j-1.

Clearly, the membership of each partitioned group G_i is maintained well during the mapping process. However, only such senders and the receiver in $\mathrm{BCube}(n, k)$ are insufficient to form an Incast tree at the level of a generalized hypercube. We need to identify a minimal set of additional servers for each stage and identify switches across successive stages to constitute an Incast tree with a low cost. Recall that a pair of neighboring servers across two successive stages connects to a common switch, whose label and level can be derived from the labels of servers. For this reason, we just focus on the selection of additional servers for each stage.

Identifying the minimal set of servers for each stage is an efficient approximate method for the minimal Incast aggregation tree problem. For any stage, fewer servers incurs fewer outgoing flows towards the receiver since all incoming flows at each server will be aggregated as a single flow. The problem is translated to derive a common neighbor at stage $j - 1$ for as many servers at stage j as possible. In this way, the number of total servers at stage $j - 1$ can be significantly reduced.

We recursively identify the minimal set of additional servers at each stage from stage $(k+1)$ to stage 1. We start with stage $k+1$, which just contains senders that are $(k+1)$-hops neighbors of the receiver. Such servers fall into groups $G_1, \ldots, G_\beta$, which are $(k+1)$ hops from the receiver. Such groups are ranked and processed in the order of their cardinalities.

For a group $|G_i| > 1$, the labels of its all servers only differ in one dimension, i.e., the partition symbol e_i. Our design insight is that all flows from G_i should be forwarded towards the common one-hop neighbor to be aggregated at stage k, called an *inter-stage aggregation scheme*. Such a common neighbor can be simply determined by e_i. It has the same label with all servers in G_i except the e_i dimension. Its label in the e_i dimension is just equal to that of the receiver's label. Such a common neighbor may appear at the group G_i or not. If yes, it has been mapped to stage k. Otherwise, it should be appended as a new server for stage k.

For a group with $|G_i| = 1$, it lacks a related partition symbol e_i. The only sender has a one-hop neighbor at stage k in each of those k dimensions. If the sender randomly selects a one-hop neighbor for stage k, the resultant Incast tree losses non-trivial aggregation opportunities. A reasonable way is that the sender selects such a one-hop neighbor for the next stage, which is just a sender of the Incast transfer if possible. Thus, the flow from group G_i and the local flow of the selected sender at the next stage can be performed by the inter-stage aggregation. In a general scenario, the inter-stage aggregation fails to be achieved for a single group when none of the possible neighbors at the next stage is a sender of the Incast transfer.

Accordingly, for a single group with $|G_i| = 1$, we design an *intra-stage aggregation scheme* to identify an aggregation server at the same stage if possible. Consider that the only sender $node_0$ in G_i at stage $k + 1$ is isolated from other groups. Fortunately, it may have a neighbor in one of other dimensions, $\{0, 1, \ldots k\}$-$\{e_i\}$, which exists in other group G_j with $|G_j| > 1$ with a partition symbol e_j. In such a case, $node_0$ no longer needs to deliver its flow towards the next stage but to such a neighbor at the same stage. The selected one-hop neighbor thus aggregates all incoming data flows and the data flow by itself at stage $k + 1$.

For the only node in a single group with $|G_i| = 1$ at stage $k + 1$, it will select the next server at stage k along with the partition symbol of the largest groups if it realizes neither inter-stage aggregation nor intra-stage aggregation. Consider that some partitioned sender groups may also appear at stage k. Thus, the set of all servers on stage k are the union of two parts, the already existing senders at stage k and all appended servers, derived from those sender groups at stage $k + 1$.

We can thus directly apply our approach to infer the required servers at the stage $k, k - 1, \ldots, 1$. However, two dedicated considerations can further reduce the cost of the resultant Incast aggregation tree. First, those already existing senders form a set of partitioned groups according to Algorithm 9.2. Thus, each appended new server should be added into a group if it is a one-hop neighbor of all senders in that group. Second, all of the rest nodes, which cannot be included in any existing groups, may have a chance to be partitioned into a few new groups according to Algorithm 9.2.

Finally, servers in all $k + 2$ stages and the directed paths across successive stages constitute an aggregation Incast tree for an uncertain Incast transfer. Such a method is called the group-based aggregation tree building method, and the resultant tree is called the group-based Incast tree.

Theorem 9.3 *For an uncertain Incast transfer with a given receiver and m non-deterministic senders in BCube(n, k), the complexity of the group-based aggregation tree building method is of $O(m^2 \times (\log N)^2)$, where $N = n^{k+1}$ is the number of servers in BCube(n, k).*

Proof In the beginning, our method has to solve the associated MSG problem via invoking Algorithm 9.2 at the cost of time complexity $O(k^2 \times m^2)$. As a result, all uncertain senders become deterministic and are partitioned into the least number of groups, irrespective of the selection of the receiver. To take a step further, the group-based method constructs an Incast aggregation tree from at most stage $k+1$ to stage 1 recursively. At each stage i $(1 \leq i \leq k + 1)$, the number of existing senders and appended new servers is definitely less than m. We try to append those isolated servers to existing groups at the cost of time complexity $O(m^2)$. Finally, we will derive a next neighbor at stage $i - 1$ for each group or each isolated server at the cost of time complexity $O(m)$. Thus, the computation cost of deriving the Incast aggregation tree is $O(k \times (m^2 + m))$ since it involves at most $k+1$ stages. The entire computation cost of our method is $O(k^2 \times m^2)$; hence, Theorem 9.3 is proved.   □

9.4 Performance Evaluation

We start with the evaluation methodology and scenarios. We then evaluate uncertain incast transfers and related work under different incast transfer sizes, data center sizes, aggregation ratio, distributions of incast members, and the receiver diversity.

9.4.1 Evaluation Methodology and Scenarios

Before evaluating the benefits of aggregating uncertain Incast transfers, we first emulate data centers of BCube(n, k) from the aspect of overlay networks. One hundred virtual machines (VMs) are hosted by ten physical servers connected together with a switched Ethernet. One VM acts as the controller node, while each of the other VMs emulates a series of virtual Bcube nodes (VBNs) on demand. Each VBN has the ability to cache, and process data flows and can support the in-network aggregation of Incast transfers.

To achieve an uncertain Incast transfer, the controller node then launches a job similar to the wordcount job built-in Hadoop. Such a job will be scheduled overall VBNs to perform a set of Map-like or Reduce-like tasks. Without loss of generality, for each file block to be processed, its replicas are selectively deployed to three VBNs.

As a result, an uncertain Incast transfer forms between those uncertain Map-like tasks and each deterministic receiver for a job. The root cause is that an uncertain map-like task for each file block can be finally scheduled to one of three related VBNs. After initializing an uncertain Incast transfer, the average flow size across the map and reduce stages is one unit (about 10 MB).

We evaluate the benefits of uncertain Incast transfers against the deterministic one by comparing four metrics of each resulting Incast aggregation trees. They are the resultant network traffic, the number of active links, the number of unique senders, and the number of partitioned groups. More precisely, we construct multiple types of Incast aggregation trees, resulting from the combination of different initialization methods and tree-building methods. The initialization methods include SD-based and MD-based MSG methods and the random non-MSG method. The tree-building methods cover interstage-based and intrastage-based building methods as well as the IRS-based method. The IRS-based Incast tree was proposed in Chap. 7 for the aggregation problem of deterministic Incast transfer.

Moreover, we regulate four essential factors to evaluate their impacts on the performance metrics. They are the Incast transfer size, the data center size, the aggregation ratio among inter-flows, and the distribution of Incast members.

9.4.2 Impact of the Incast Transfer Size

A MapReduce-like job often involves several hundred or thousands of Map tasks. To measure the scaling properties of the methods proposed in this chapter, we deploy a series of uncertain Incast transfers over BCube(8,4). For each uncertain Incast transfer, the number of uncertain senders ranges from 100 to 4000 with the interval 100, while the receiver is fixed.

We collect the amount of network traffic and the number of active links after completing each uncertain Incast transfer along with different Incast aggregation trees. We plot the changing trends of the performance metrics under the three tree-building methods while the same MD-based initialization method in Fig. 9.3. The intrastage-based and interstage-based Incast trees for an uncertain Incast transfer can save the caused network traffic up to 33.85 and 27% on average than the prior IRS-based Incast tree. It is clear that the intrastage-based tree building method considerably outperforms the other two building methods. Note that similar evaluation results under the SD-based initialization method are achieved.

Based on the above results, the intrastage-based tree building method should be preferred among the three building methods. We then compare the performance metrics under the varied initialization methods but the same intrastage-based tree building method. Figure 9.4a, b show that the MD-based and SD-based methods outperform the random initialization method due to incur less amount of network traffic and occupy fewer numbers of links.

The benefits of MD-based and SD-based methods are further evidenced by Fig. 9.4c, d. First of all, the two MSG-based methods utilize fewer unique senders

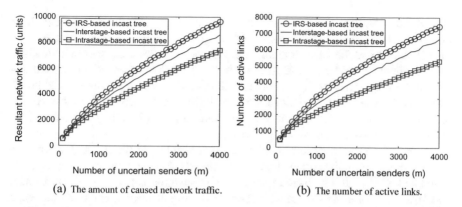

(a) The amount of caused network traffic.

(b) The number of active links.

Fig. 9.3 For uncertain Incast transfers in BCube(8,4), the changing trends of two metrics with flow amounts

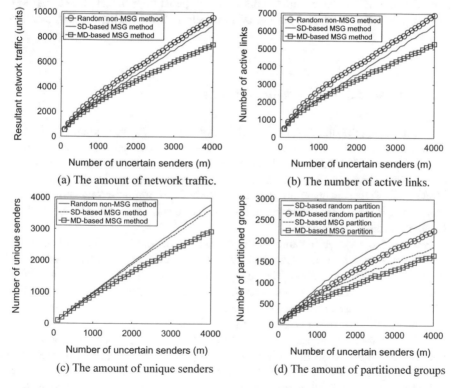

(a) The amount of network traffic.

(b) The number of active links.

(c) The amount of unique senders

(d) The amount of partitioned groups

Fig. 9.4 For uncertain incast transfers in BCube(8,4), the changing trends of four metrics under different initialization methods

than the random initialization method. Actually, the two methods consider those potential senders shared by some flows of the give uncertain Incast transfer. Second, the MD-based and SD-based initialization methods also naturally partition the given set of senders into fewer groups. As a result, more flows of an uncertain Incast transfer will be locally aggregated at some senders or at one-hop neighbors, and hence significantly decrease the resultant network traffic. Although similar ideas can be used to partition those senders after the random initialization, they will bring more partitioned groups, as shown in Fig. 9.4d.

Clearly, with the same intrastage-based tree building method, the performance under different initialization methods varies greatly. More precisely, using the MD-based and SD-based initialization methods can save the network traffic up to 22.6 and 16%, compared with the random non-MSG initialization method. At the same time, the number of active links can be reduced up to 29 and 20%, respectively. Less active links mean that the Incast tree occupies fewer servers and network devices. Note that the MD-based method achieves more gains than the SD-based method. Thus, the MD-based method should be the preferred one during the initialization process.

9.4.3 Impact of the Data Center Size

After comparing the different Incast tree-building methods, we evaluate the impact of the data center size on the performance of the resulting Incast tree. We emulate BCube(8,k), $k \in \{3, 4, 5, 6, 7\}$ and generate the uncertain Incast transfer with $m = 1000$ uncertain senders and any given receiver, which are randomly deployed in BCube(8,k). The given uncertain Incast transfer will be complete along with different Incast trees, formed by the preferred intrastage-based building method under the three different initialization methods.

Figure 9.5 shows that the amount of network traffic and active links of the three Incast trees always grow up along the increase of the data center size, irrespective of the used initialization method. The MD-based method exhibits remarkable advantages over than other two methods for middle size data centers when $k < 5$, while it has similar performance when $k \geq 5$. Note that BCube(8,5) refers to a large-scale data center with 262144 servers. The fundamental reason is that the benefit of the two MSG methods becomes weak due to the extremely sparse distribution of all members of an uncertain Incast transfer in huge-scale data centers. Figure 9.5c, d provide additional evidences for such an observation. The number of unique senders under the three initialization methods exhibit trivial differences after k exceeds 5, and the number of partitioned sender groups increases.

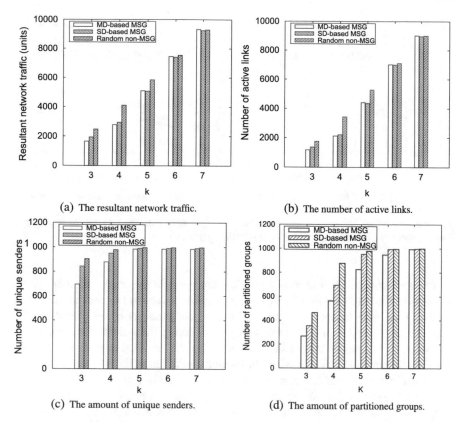

(a) The resultant network traffic.

(b) The number of active links.

(c) The amount of unique senders.

(d) The amount of partitioned groups.

Fig. 9.5 The changing trends of four metrics with data center size under different initialization methods

9.4.4 Impact of the Aggregation Ratio

To ease the explanation, the prior analysis assumes that the size of new flow after aggregating multiple flows, consisting of key-value pairs, equals that of the largest flow. The evaluations indicate that it is not necessarily always true. For this reason, we evaluate the performance of tree-building methods under different aggregation ratios.

Given s data flows in a general uncertain Incast transfer, let f_i denote the size of the ith data flow for $1 \le i \le s$. Let δ denote the aggregation ratio among any number of data flows, where $0 \le \delta \le 1$. Thus, the size of the resultant new data flow after aggregation is given by

$$\max\{f_1, f_2, \ldots, f_s\} + \delta \times \Big(\sum_{i=1}^{s} f_i - \max\{f_1, f_2, \ldots, f_s\}\Big) \qquad (9.1)$$

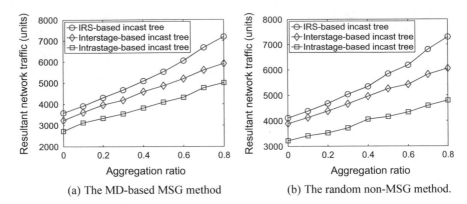

(a) The MD-based MSG method   (b) The random non-MSG method.

Fig. 9.6 The changing trends of caused network traffic along with the increase of aggregation factors, under two initialization and three tree building methods

Previous analysis just tackles a special scenario where $\delta = 0$. Another special scenario is that each key in one of such s data flows does not appear at other data flows, i.e., $\delta = 1$. In such a special case, an initialized uncertain Incast transfer cannot achieve any gain from the aggregation among the s data flows. Such two special scenarios rarely happen in practice. Therefore, we further evaluate the three tree-building methods under the MD-based and random initialization methods for an uncertain Incast transfer in a more general scenario where $0 \leq \delta \leq 0.8$.

Figure 9.6 shows that the preferred intrastage-based tree building method always outperforms the other two methods when $\delta \leq 0.8$. Additionally, we can see that the MD-based initialization always outperforms the random initialization method. Assume that the value of the random variable δ follows a uniform distribution (if any). In this setting, the intrastage-based Incast tree saves the network traffic by 35 and 33% under two initialization methods, compared to the IRS-based method. Thus, our intrastage-based building method considerably outperforms other methods for general uncertain Incast transfers, irrespective of the used initialization methods.

9.4.5   Impact of Distributions of Incast Members

Currently, all members of an uncertain Incast transfer are scheduled to idle servers in the scope of the entire data center. That is, the senders and the receiver are somehow randomly distributed in the data center. Such a random distribution makes the uncertain Incast transfer occupy more data center resources and generate more network traffic. This section extends our MD-based initialization method and intrastage-based tree building method to involve uncertain Incast transfers, whose members are distributed in a managed manner.

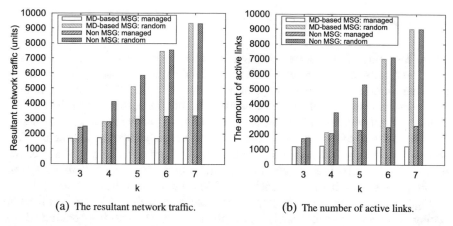

(a) The resultant network traffic.   (b) The number of active links.

Fig. 9.7 The performance metrics under two distributions of Incast transfer members

The key insight is to find the smallest subnet $BCube(n, k_1)$ in $BCube(n, k)$ such that all members of an uncertain Incast transfer satisfy their schedule constraints in the subnet. In such a case, the hamming distance between the receiver and each potential sender of the Incast transfer is at most $k_1 + 1$, i.e., their labels just differ in at most $k_1 + 1$ dimensions. The intrastage-based Incast tree in the scope of $BCube(n, k_1)$ for such an Incast transfer has at most k_1+1 stages. In theory, it occupies fewer data center resources and incurs less network traffic than the prior Incast tree in the entire scope of $BCube(n, k)$.

We generate a series of uncertain Incast transfers, each of which consists of 1000 uncertain senders under $BCube(8,k)$ for $3 \le k \le 7$. For each setting of k, we apply the two distribution schemes (i.e., managed distribution and random distribution schemes) to all its Incast transfers, under the same tree-building method but different initialization methods (i.e., MD-based and random initialization methods). Figure 9.7 plots the network traffic and active links of each uncertain transfer under two different deployment schemes. The two metrics of Incast trees under the managed distribution are relatively stable, and always less than that in random distribution, irrespective of the initialization method. However, the Incast trees with the random distribution scheme exhibit increasing amount of network traffic and active links as the increase of k. That is, the Incast trees with the MD-based and random initialization exhibit a very similar performance in huge-scale data centers after k exceed 5.

The tree-building approach in the managed distribution scheme saves the network traffic by 53 and 42% compared to that in the distribution scheme under the MD-based and random initialization methods, respectively. Such results indicate that our tree-building methods and initialization methods can achieve more gains in the managed distribution scheme.

9.4.6   Impact of the Receiver Diversity

When designing the Incast tree-building methods in previous sections, we do not focus on picking the best receiver due to the following two reasons: (1) Too many potential receivers bring huge computation overhead to pick the best one, given the set of deterministic senders. (2) The Incast tree-building methods proposed in this chapter are not very sensitive to receiver selection. To verify the second reason, we carry out two sets of experiments. In the first set, we study uncertain Incast transfers with $m = 100$, $m = 500$, and $m = 1000$, each of which has 100 candidates of the receiver. In the second set, we consider uncertain Incast transfers with $m = 2000$, $m = 3000$, and $m = 4000$, each of which has 1000 candidates of the receiver.

Figure 9.8 plots the changing trend of the caused network traffic under different settings of uncertain Incast transfers. It is clear that those Incast trees exhibit trivial differences in the amount of network traffic. Table 9.1 shows the mean and standard deviation of the amount of caused traffic under varied uncertain Incast transfers. The standard deviation is always small even the amount of uncertain senders increases up to 4000. Accordingly, we infer that our initialization and tree-building methods can well address the problem of receiver diversity. That is, we can select a receiver randomly.

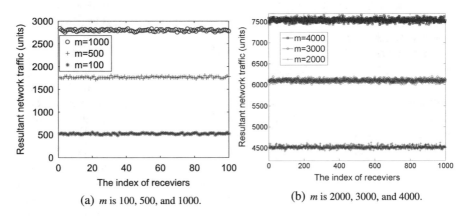

(a) m is 100, 500, and 1000.

(b) m is 2000, 3000, and 4000.

Fig. 9.8 For uncertain Incast transfers, the changing trend of network traffic, resulting from Incast trees under different receivers, is relative smooth

Table 9.1 The mean and standard deviation of the amount of caused traffic under varied uncertain Incast transfers

	$m = 100$	$m = 500$	$m = 1000$	$m = 2000$	$m = 3000$	$m = 4000$
μ	521.3	1764.7	2796.3	4518.2	6101.8	7535.7
σ	13.23	18.8	19.19	31.05	32.87	41.62

References

1. Chowdhury M, Zaharia M, Ma J, et al. Managing data transfers in computer clusters with orchestra [J]. ACM SIGCOMM Computer Communication Review, 2011, 41(4): 98–109.
2. Guo D, Xie J, Zhou X, et al. Exploiting Efficient and Scalable Shuffle Transfers in Future Data Center Networks [J]. EEE Transactions on Parallel and Distributed Systems, 2015, 26(4): 997–1009.
3. Abts D, Marty M R, Wells P M, et al. Energy proportional datacenter networks [C]. In Proc. of 37th ACM ISCA, Saint-Malo, France, 2010, 338–347.
4. Ahn J H, Binkert N, Davis A, et al. HyperX: topology, routing, and packaging of efficient large-scale networks [C]. In Proc. of ACM/IEEE SC, Portland, Oregon, USA, 2009.

Chapter 10
Collaborative Management of Correlated Multicast Transfer

Abstract Multicast is designed to jointly deliver the same content from a single source to a set of destinations; hence, it can efficiently save the bandwidth consumption and reduce the load on the source. Distributed file systems in data centers provide multiple replicas for each data block. In this case, the traditional Multicast faces the diversity of senders since it is sufficient for each receiver to get a replica from any sender. This brings new opportunities and challenges to reduce the bandwidth consumption of a multicast transfer. This chapter focuses on such Multicast with uncertain senders and constructs an efficient routing forest with the minimum cost (MCF). MCF spans each destination by one and only one source, while minimizing the total cost (i.e. the weight sum of all links in one multicast routing) for delivering the same content from the source side to all destinations. Prior approaches for deterministic Multicast do not exploit the opportunities of a collection of sources; hence, they remain inapplicable to the MCF problem. Actually, the MCF problem for a multi-source Multicast is proved to be NP-hard. Therefore, we propose two $(2 + \varepsilon)$-approximation methods, named P-MCF and E-MCF. We conduct experiments on our SDN testbed together with large-scale simulations under the random SDN network, regular SDN network and scale-free SDN network. All manifest that our MCF approach always occupies fewer network links and incurs less network cost for any uncertain Multicast than the traditional Steiner minimum tree (SMT) of any related deterministic Multicast, irrespective of the used network topology and the setting of Multicast transfers.

10.1 Introduction

Delivering the same content from a single source to a group of destinations separately usually occupies excessive bandwidth since the source needs to achieve a Unicast transfer with each destination. To solve this problem, academia and industry propose many Multicast protocols, which save bandwidth consumption and reduce the load on the sender. The root cause is that Multicast can avoid unnecessarily duplicated transmissions among a set of independent Unicast paths towards destinations after multiplexing a shared Multicast tree. Despite such bandwidth superiority, Multicast

© Springer Nature Singapore Pte Ltd. 2022
D. Guo, *Data Center Networking*,
https://doi.org/10.1007/978-981-16-9368-7_10

on the Internet has suffered many deploying obstacles during the past decades. Until recently, it achieves a few successful network-level deployments in IPTV networks [1], enterprise networks, and data center networks [2–4].

Software-defined networking (SDN) [5] is a flexible architecture for network resource management and network application innovation. After separating the control plane and data plane, the controller is responsible for delivering a programmable control plane. It collects users' network requirements and jointly manages the underlying physical network in a centralized way. The response to users' requirements will be translated as a series of control policies delivered to involved network devices to update their flow tables. Accordingly, SDN provides an excellent opportunity for deploying various flexible protocols, such as traffic engineering, energy-saving, Multicast protocols, etc. In this setting, Multicast for SDN still attracts much less attention from both academia and industry.

Among existing Multicast routing protocols, PIM (protocol independent Multicast) is the most widely used one [2], which connects the source and destination via a shortest-path tree. An inherent drawback of this method is that those independent shortest paths, from each destination to the same receiver, lack opportunities to share more common links. Thus, the formed shortest-path tree fails to minimize the number of used links; hence, Multicast traffic along it cannot considerably save the bandwidth consumption. For this reason, many efforts focus on solving the Steiner minimum tree (SMT) [6], an NP-hard problem. It aims to minimize the total number of edges in a tree, spanning all members of any given Multicast group. The SMT performs better than PIM; however, it is not adopted by current networks due to the huge computation overhead even under heuristic methods and the challenge of distributed deployment [7]. Fortunately, the emergence of SDN makes it possible to realize novel Multicast protocols.

However, such Multicast methods assume prior knowledge of Multicast characteristics, i.e., the source of each Multicast group is fixed in advance. We call this case the *deterministic Multicast*. Unfortunately, in many cases, the characteristics of a Multicast transfer are unknown a priori. It is not necessary that the source of a Multicast transfer has to be placed in a specific location as long as certain constraints are satisfied. A major reason is a widely used content replica design for improving the robustness and efficiency in various networks [8]. When delivering a content file to multiple destinations, the source of such a Multicast transfer can be anyone replica in theory. This causes the *uncertain Multicast* due to the source flexibility.

In this chapter, we focus on a new type of multicast, named the uncertain Multicast. Given destinations and uncertain sources, together with the network topology, the uncertain Multicast problem aims to construct a forest. It ensures that each destination connects to just one source through a path in the forest. The proposed uncertain Multicast is a general scheme, which is equivalent to the deterministic Multicast in the special setting of a single source. The objective is to propose the construction methods of Multicast forest for uncertain Multicast and further minimize the transmission cost of this uncertain Multicast. Compared with the SMT, finding the minimum cost forest (MCF) for an uncertain Multicast is more challenging due to the flexible usage of multiple sources and the impact on routing. We prove that the MCF problem is NP-hard. A potential approach is to divide an uncertain Multicast as a set of deterministic Multicasts, each with a distinct source. The SMT with the least cost among such Multicasts just acts as the MCF of the uncertain Multicast. This way, however, suffers the complexity of solving a set of NP-hard SMT problems. Moreover, the MCF under non-single sources may cause less total cost than the picked best SMT with any single source. Thus, prior approaches, relying on traditional Multicast, remain inapplicable to the uncertain Multicast proposed in this chapter.

To solve the MCF problem efficiently, we propose two $(2 + \varepsilon)$-approximation methods, P-MCF and E-MCF, which can be deployed in SDN controllers. Such methods use a part of sources to establish routing paths towards all destinations. They can seek shared nodes among such paths to enhance the possibility of aggregating more links, reducing the total cost of resultant MCF. We conduct small-scale experiments on our SDN testbed and large-scale simulations to compare the uncertain Multicast and traditional Multicast under the random network, regular network and scale-free network, respectively. The evaluation results indicate that the MCF of an uncertain Multicast occupies fewer network links and incurs less network cost, irrespective of the used network topology.

10.2 Related Work

Given any Multicast group, many research work focuses on constructing Multicast trees to satisfy various constraints. The SMT problem formulates a common constraint on the Multicast tree, i.e., the number of occupied links spanning all members of a Multicast group. Many algorithms for the SMT problem have been proposed to approximate the optimal solution. A fast and effective algorithm achieves a $(2 + \varepsilon)$-approximation, which approximates the SMT problem using a minimum spanning tree (MST) in Literature [9]. Several methods, based on greedy strategies achieve better approximation ratio, such as 1.746 [10], 1.693 [11], 1.55 [12], but considerably incur high time complexity than the $(2 + \varepsilon)$-approximation algorithm. The algorithm proposed in Literature [12] achieves a 1.55-approximation. However, the time complexity is too high due to many rounds of iterative computation, which is hard to be terminated within the acceptable time. Actually, the round of iterations is unpredictable for large-scale networks. Although the fast SMT algorithm does not achieve the optimal approximation ratio, it significantly saves more computation time than other algorithms. Hence, it is more suitable than other algorithms for large Multicast groups and large-scale networks. HOWEVER, such SMT algorithms are still inapplicable to the MCF problem proposed in this chapter due to the introduction of multiple sources for each Multicast group.

Current networks have not adopted those Multicast algorithms due to the huge computation overhead, and the challenge of distributed deployment [7]. Fortunately, the emergence of SDN makes it possible to realize novel Multicast protocols. For example, Literature [13] deploys a network virtualization application on SDN control planes, simplifying the unified management of Multicast requirements and Multicast tree. In addition, Literature [13] proposes a reliable Multicast routing algorithm in SDN, namely, RAERA. Given any Multicast group, it can construct a reliable Multicast tree with the shortest paths. In this way, each receiver can find at least one reliable recovery node from original receiving paths to obtain data when Multicast transfer fails. RAERA constructs a reliable Multicast tree, significantly improving the reliability of existing Multicast protocols.

The most related research work revolves around the multi-source Multicast in Literature [15], which delivers a stream from multiple sources to a group of destinations. The MMForest algorithm can establish a Multicast forest to span all destinations and sources under the constraint of achieving the maximal residual bandwidth for each Multicast stream. Actually, MMForest is designed based on the existing Widest-Path Forest algorithm, which prefers to select those links with higher residual capacity. It means that different Multicast streams have to design their MMForests cooperatively.

The MMForest method focuses on the design constraint of optimizing the residual bandwidth under a set of Multicast streams. On the contrary, the MCF problem proposed in this chapter aims to tackle the widely used constraint, i.e., minimizing the network resources assumed by each Multicast stream. Due to utilizing the minimal number of links and assuming as little network bandwidth as possible, the MCF problem can further improve the level of residual bandwidth when more Multicast streams share the network concurrently. Moreover, the proposed MMForest exhibits poor performance when addressing our MCF problem. The root cause is that the basic idea is similar to the most straightforward Multicast tree. Each destination selects one path with the largest residual bandwidth towards all sources and then merges those selected paths from all destinations. Obviously, this method loses non-trivial opportunities to maximize the number of shared links among individual paths.

10.3 Problem Statement of Uncertain Multicast

We start with the important observation about uncertain Multicast and then present the problem statement of uncertain Multicast. Finally, we give a mixed-integer linear programming model for the uncertain Multicast problem.

10.3.1 Observations

Multicast is a natural approach to deliver the same content to a group of destinations, not a single destination. It benefits in efficiently saving the bandwidth consumption and reducing the load on the sender, compared with Unicast. The total cost of a Multicast tree, which spans all destinations and the single source, is widely used as the performance metric of a Multicast. Without considering the diversity among network links, the cost of Multicast tree mostly depends on the number of occupied links. To

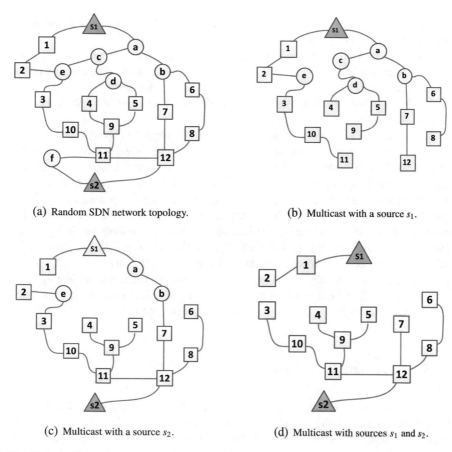

(a) Random SDN network topology.

(b) Multicast with a source s_1.

(c) Multicast with a source s_2.

(d) Multicast with sources s_1 and s_2.

Fig. 10.1 An illustrative example of an uncertain multicast with the same destination set

minimize the tree cost, given a Multicast group, many approximation methods [16] have been proposed to construct the SMT.

As aforementioned, the source of a Multicast transfer is usually unfixed. Any replica can serve as a single source because of the widespread use of content replica designs in data centers. Given the same set of destinations, the cost of a Multicast group is sensitive to the selection of sources. Different selections of sources result in diversity SMT and total link cost.

Figure 10.1 plots an illustrative example of the diversity of SMT for Multicast with different sources and the same destinations. Figure 10.1a is a random network topology, where the triangle nodes denote possible sources, the square nodes denote the destinations, and the cycle nodes denote intermediate forwarding nodes. Figure 10.1b, c indicate two Multicast trees with s_1 and s_2 as the single source, respectively. We can see that the Multicast tree with the source s_1 is of cost 17, while the

Multicast tree with the source s_2 is of cost 16, even node s_1 also participates in the Multicast routing as a forwarding node.

Such facts motivate us to resolve the problem of Multicast with uncertain sources, i.e., the uncertain Multicast problem. As described above, an uncertain Multicast can be initialized as a series of deterministic Multicasts with different sources. We can further build a SMT for each deterministic Multicast, and the SMT with the least tree cost should be the preferred one.

The source of a Multicast is not necessary to be placed in specific locations as long as certain constraints are satisfied. A major reason is the widely used design of content replica designs for improving the robustness and efficiency in various networks. Thus, any replica can participate in the Multicast transmission by acting as the single source, and the Multicast routing is a group of trees instead of a single tree. Different selections of sources result in different Multicast trees and costs. For example, the Multicast tree in Fig. 10.1c is a more desired one than that in Fig. 10.1b. However, we find that the picked best SMT with any single source may not be the optimal selection. A more efficient Multicast may exist when multiple sources participate simultaneously. For example, the Multicast transmission in Fig. 10.1d is just of cost 12 and outperforms any individual SMT. However, such a preferred design cannot be achieved by prior approaches, relying on deterministic Multicast.

10.3.2   Problem Statement

In uncertain Multicast, a destination can get data from each potential source rather than a specific one. Thus, it has the opportunity to further reduce the transmission cost. This chapter aims to fully exploit the benefits of multiple sources and minimize the transmission cost of uncertain Multicast.

More precisely, consider a network $G(V, E)$, where V and E denote the set of nodes (switches and servers) and edges (links), respectively. Each edge $e \in E$ is associated with a cost. The cost of an uncertain Multicast is calculated as the total cost of its occupied links. We formalize the uncertain Multicast problem in Definition 10.1.

Definition 10.1 Given a destination set $D \in V$ and a potential source set $S \in V$, an uncertain Multicast means to deliver the same content to all destinations in the set D from partial even all sources in the set S. A constraint is that each destination just reaches one and only one source in set S.

The uncertain Multicast is a general scheme of Multicast communication. Actually, it is equivalent to the deterministic Multicast when the set S contains only one source.

However, Definition 10.1 considers no constraint on how to select source for each destination $d \in D$. Thus, even in the case of $|S| > 1$, an uncertain Multicast will be a deterministic one when all destinations employ any $s \in S$ as their common source. It is clear that the usage strategy of such sources dominates the transmission cost of an uncertain Multicast. Therefore, we present the minimum cost forest (MCF) problem in Definition 10.2 to minimize the transmission cost of an uncertain Multicast.

Definition 10.2 Given an uncertain Multicast, the problem of MCF is to find a forest satisfying the following two constraints. First, the forest spans each destination $d \in D$ with only one source $s \in S$. Second, the total cost of utilized edges in the forest is minimized.

Accordingly, we can infer that the MCF does not necessarily contain all sources. Any pair of sources s_1 and s_2, however, should be separated if they were connected in the MCF. Otherwise, some destinations may reach more than one source, increasing the cost of the resultant forest. The Multicast topology in Fig. 10.1c is not desirable since two sources s_1 and s_2 are linked through a path, which introducing unnecessary links into MCF and increasing the cost. That is, in the MCF containing α sources which belong to S, there should be exactly α isolated trees, each of which using one of the α sources as its root.

The uncertain Multicast becomes the deterministic Multicast when the source set contains only one element. Thus, the MCF problem is equivalent to constructing an SMT, which is NP-hard in a general graph. In a general setting, the MCF problem involves more challenges than the SMT problem due to the flexible usage of multiple sources.

Theorem 10.1 *Given an uncertain Multicast with the source set S and destination set D in a network $G(V, E)$, the problem of calculating its MCF is NP-Hard.*

Proof We prove the NP-hardness of MCF by giving a polynomial-time reduction from SMT, an NP-hard problem, to MCF. We first consider the SMT problem of an uncertain Multicast. It means to find an SMT for spanning all destinations and sources of that uncertain Multicast. Clearly, the optimal solution to this SMT problem cannot be found within polynomial time. Note that $|S|^2$ source pairs exist in the optimal SMT, and each source pair is connected by only one path. The path between any two sources has at least one redundant link. A link is redundant only if each destination still reaches at least one source after removing that link. For example, links $a \rightarrow b$ and $s_1 \rightarrow a$ in Fig. 10.1c are redundant. Therefore, the SMT problem of an uncertain Multicast would be reduced to the MCF problem by removing all redundant links in the optimal SMT. The process of removing potential redundant links should be completed within polynomial time. The remaining links in the original optimal SMT form a forest F, where each destination connects to only one source. Thus, the MCF problem of an uncertain Multicast is NP-Hard; hence, Theorem 10.1 is proved.   □

10.3.3   Mixed Integer Linear Programming

We further present a Mixed Integer Linear Programming (MILP) formulation for the proposed MCF problem.

Let N_v denote the set of all neighbor nodes of node v in G, and u is in N_v if $e_{u,v}$ is an edge from u to v. Let S denote the source set of the uncertain Multicast group, and D denote the destination set. The output minimal cost forest F needs to ensure that there is only one path in F from every node $d \in D$ to only one source $s \in S$. To achieve this goal, our problem includes the following binary decision variables. Let binary variable $\omega_{d,s}$ denote whether a destination node d selects a source node s as the root node. That is, there is a path from d to s if $\omega_{d,s} = 1$. In this setting, let binary variable $\pi_{d,u,v}$ denote whether edge $e_{u,v}$ is in the path from the destination d to the source s. Let binary variable $\varepsilon_{u,v}$ denote whether edge $e_{u,v}$ is in the output forest F. Intuitively, we should find the path from each destination d to just one source node s with $\omega_{d,s} = 1$. Thus, every edge $e_{u,v}$ in the path has $\pi_{d,u,v} = 1$. The routing of the resultant forest with $\varepsilon_{u,v} = 1$ for every edge $e_{u,v}$ in F can be achieved by the union of the paths from all destination nodes in D to at least one source node in S.

The objective function of the MCF problem is as follows.

$$\min \sum_{e_{u,v} \in E} c_{u,v} \times \varepsilon_{u,v},$$

where the weight $c_{u,v}$ denotes the cost of edge $e_{u,v}$.

To find $\varepsilon_{u,v}$, our MILP formulation includes the following constraints, which explicitly describe the routing principles for the uncertain Multicast.

$$\sum_{s \in S} \omega_{d,s} = 1, \forall d \in D \tag{10.1}$$

$$\sum_{v \in N_s} \pi_{d,s,v} = 1, \omega_{d,s} = 1, \forall d \in D, \exists s \in S \tag{10.2}$$

$$\sum_{u \in N_d} \pi_{d,u,d} = 1, \omega_{d,s} = 1, \forall d \in D, \exists s \in S \tag{10.3}$$

$$\sum_{v \in N_u} \pi_{d,u,v} = \sum_{v \in N_u} \pi_{d,v,u} = 1, \quad \omega_{d,s} = 1,$$
$$\forall d \in D, \exists s \in S, \forall u \in V, u \neq d, u \neq s \tag{10.4}$$

$$\pi_{d,u,v} \leq \varepsilon_{u,v}, \forall d \in D, \exists s \in S, \forall e_{u,v} \in E \tag{10.5}$$

For each destination node $d \in D$, the first constraint ensures that there exists only one source node $s \in S$ such that there exists at least one path from d to s in the output F. Formulas (10.2), (10.3), and (10.4) impose constraints on finding the path from every destination d in D to its source s. More precisely, given any destination node d, s is the source of the path towards d, and Formula (10.2) implies that only one edge $e_{s,v}$ from s to any neighbor node v needs to be selected with $\pi_{d,s,v} = 1$. At the same time, it ensures that there exists only one path from s to d and any pair of source nodes in the output F is isolated. On the other hand, every destination node d is the flow destination, and Formula (10.3) ensures that only one edge $e_{u,d}$ from any neighbor node u to d must be selected with $\pi_{d,u,d} = 1$. Furthermore, for any other node u in G, we can infer from Formula (10.4) that it is either located in the path from s to d or not. If u locates in the path, u has one incoming flow in the path with one binary variable $\pi_{d,v,u} = 1$ and one outgoing flow in the path with one binary variable $\pi_{d,u,v} = 1$. Otherwise, $\pi_{d,v,u}$ as well as $\pi_{d,u,v}$ are 0. Note that, according to the objective function, $\pi_{d,v,u} = 1$ is set for just one neighbor node v so as to minimize the cost of output forest F. Formulas (10.5) desires to find the routing of the output F, i.e., $\varepsilon_{u,v}$. More specifically, $\varepsilon_{u,v}$ must be 1 if edge $e_{u,v}$ is in the path between at least one pair of source s and destination d, i.e., $\pi_{d,u,v} = 1$. The output forest F is the union of the paths from all destinations to their corresponding sources.

10.4 Efficient Building Methods of MCF

We start with designing a fundamental approximation method, P-MCF, to find the minimum cost forest and then present a more efficient method, E-MCF.

10.4.1 Primary Approximation Method

As aforementioned, the MCF problem for an uncertain Multicast is NP-hard and cannot be solved in polynomial time. Thus, we focus on designing efficient methods to approximate the optimal forest [17] for the uncertain Multicast.

A straightforward method is to treat an uncertain Multicast as a set of deterministic Multicasts, each with a different source. Accordingly, we select the SMT with the least cost as the MCF of the uncertain Multicast. This method suffers the complexity of solving a set of NP-hard SMT problems. Moreover, the MCF involving multiple sources may cause less total cost than the picked best SMT. Thus, previous approaches for traditional Multicast remain inapplicable to the proposed uncertain Multicast. For this reason, we design an approximation MCF method, called P-MCF, as illustrated in Algorithm 10.1.

Given a network, its undirected graph model is $G = (V, E)$. For an uncertain Multicast with the source set S and destination set D, we derive the complete graph $G_1 = (V_1, E_1)$ from the original graph D and two nodes sets, S and D. The node set V_1 is the union of source set S and destination set D. According to the definition of

---

Algorithm 10.1 P-MCF( )

Require: An undirected graph $G = (V, E)$, the set of destinations D, and the set of sources S.
Ensure: A minimum cost forest F.
1: Construct the complete graph $G_1 = (V_1, E_1)$ from G and two sets, D and S.
2: Find a minimum spanning tree T for the graph G_1.
3: Construct the subgraph T_1 from T by deleting some edges, if necessary, so as to isolate all source nodes.
4: Construct the subgraph F of G by replacing each edge in T_1 by its corresponding shortest path in G. Note that those isolated source nodes in T_1 are also removed.

the complete graph, there exists an edge between any pair of nodes in V_1. For every edge $\{v_i, v_j\} \in E_1$, $d(\{v_i, v_j\})$ denotes its cost, i.e., the length of the shortest path from node v_i to node v_j in the original graph G. That is, each edge in G_1 refers a shortest path in G.

We then construct a minimum spanning tree T_1 for the graph $G_1 = (V_1, E_1)$ such that any pair of nodes in V_1 is connected through only one path in T. Accordingly, any pair of source nodes in S is also connected in the tree T. However, any two sources should be separated if they are employed in the MCF, as discussed in the previous section. For this reason, any pair of connected sources in T need to be separated by removing the maximum-cost edge along their only path in T. If there are several such edges in the path, it is reasonable to pick an arbitrary one. The challenging issue is the processing order of those connected sources in the tree T. Thus, we record the maximum-cost edge in each pair of connected source nodes' paths and sort such edges in the descending order of their costs. We remove the first edge from T and update the sorted edges. Cutting edges are finished until the updated T_1 no longer contains any connected sources. Finally, we can build an MCF for the uncertain Multicast by removing those isolated nodes in T_1 and replacing each edge in T_1 by its corresponding shortest path in G.

Figure 10.2a illustrates an example of uncertain Multicast in a small scale network. Let $S = \{s_1, s_2, s_3\}$ be the set of source nodes and $D = \{1, 2, 3, 4, 5, 6, 7\}$ be the set of destination nodes. Figure 10.2b shows a minimal spanning tree T of the complete graph consisting of all sources and destinations. Figure 10.2c shows the minimum spanning tree T_1 after removing edges $\{s_2, 2\}$ and $\{s_3, 7\}$ from T. Note that we may remove different edges to ensure that any two sources are disconnected. For example, it's reasonable to deleting edges $\{s_1, 6\}$ and $\{s_1, 1\}$. Figure 10.2d shows the resultant MCF of this uncertain Multicast with 11 links.

Theorem 10.2 *Given any uncertain Multicast in a network G, the output MCF of our P-MCF method obtains an approximation ratio of $(2 + \varepsilon)$.*

Proof Let a node set P denote the union of the source set S and the destination set D. Let $T(P)$ denote the generated MCF of the uncertain Multicast through our P-MCF method. $T(P)$ utilizes less links than the minimum spanning tree $mst(P)$, since $T(P)$ is derived from $mst(P)$ after removing some edges. We can infer that the cost of $T(P)$ is less than that of $mst(P)$, i.e., $|T(P)| \leq |mst(P)|$. Let $smt(P)$ denote a Steiner minimum tree spanning all nodes in P, which uses less links than $mst(P)$ because of link aggregation [12]. There exists an Euler tour of $smt(P)$, called T_E, which passes each edge in $smt(P)$ two times. The distance function of the Steiner tree problem conforms to the triangle inequality. The cost of any Euler tour is less than that of the related minimum spanning tree. Hence, $2|smt(P)| \geq |T_E| \geq |mst(P)| \geq |T(P)|$ [18]. Let opt be the optimal MCF of the uncertain Multicast. If additional d edges are added into opt to connect each trees as a new tree T', T' utilizes number of $|opt| + d$ links, which is more than $smt(P)$. Thus, we have $2(opt + d) \geq 2smt(P) \geq mst(P) \geq T(P)$. The approximation ratio of our P-MCF is $\frac{T(P)}{opt} \leq 2 + \frac{2d}{opt}$. Specially, the ratio of d to opt is ε and is considerably less than 1 in most real networks. Thus, Theorem 10.2 is proved.

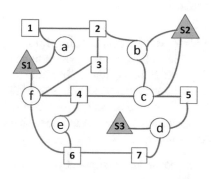

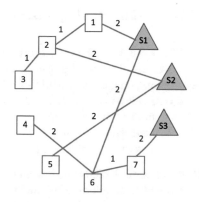

(a) An uncertain Multicast in a small-scale random network.

(b) T, the minimal spanning tree of G_1.

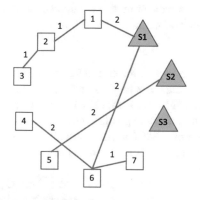

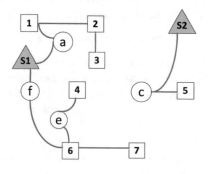

(c) T_1, where sources are disconnected.

(d) The MCF of cost 11.

Fig. 10.2 An illustrative example of our P-MCF method

We analyze the time complexity of the P-MCF method in the following way. At the beginning, number of $(|S| + |D|)^2$ shortest paths have to be calculated for achieving the complete graph G_1. The time complexity of calculating one shortest path is $O(V^2)$ [18]. Thus, the time complexity of the complete graph is $(|S| + |D|)^2 \times |V|^2$ [19]. The time complexity of finding the minimal spanning tree T from G_1 is $O((|S| + |D|)^2)$ [19]. The time complexity of finding and removing the maximum cost edge in paths between sources will cost at most $O(|S|)$. Thus, the total time complexity of the proposed P-MCF method is $(|S| + |D|)^2 \times |V|^2$.

10.4.2   Enhanced Approximation Method

The P-MCF method performs well in building MCF for uncertain Multicast; however, it still can be improved and enhanced. In this section, we design an enhanced approximation method, called E-MCF, to build the MCF for any uncertain Multicast. The basic idea is originated from the observation about the shared nodes in MCF. The major difference between the E-MCF and P-MCF methods is the construction of the complete graph G_1 from G and the uncertain Multicast. The two methods share the same process after deriving out the complete graph G_1.

Definition 10.3   Those nodes, which frequently appear in the shortest paths from sources to destinations, are called the shared nodes.

In the E-MCF method, the complete graph G_1 contains not only nodes in sets D and S but also some shared nodes. Those shared nodes are helpful to find which links are more beneficial if they are involved in the constructed minimum spanning tree T. Without such shared nodes, T usually involves more edges, which cannot be aggregated along paths from destinations to such sources.

1. **Observation of shared nodes**

Note that, in the case of the P-MCF method, all nodes in the complete graph G_1 are just those source nodes and destination nodes. Thus, it's hard to distinguish which edges/paths in G_1 are more possible to aggregate if they appear in the minimum spanning tree T. We find that if some shared nodes are appended into the complete graph G_1, the formed minimal spanning tree and final MCF are dramatically changed since more paths can be aggregated. As a result, the total transmission cost of an uncertain Multicast can be considerably reduced.

For example, as shown in Fig. 10.3, we add shared nodes f and c into G_1. Figure 10.3a reports the resultant minimum spanning tree of G_1, which differs with the tree in Fig. 10.2b apparently. The three sources are connected in this minimum spanning tree. After removing partial links to isolate the three sources, we get the

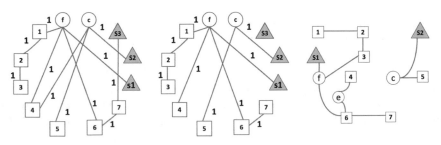

(a) T, the minimal spanning   (b) T_1, where sources are dis-   (c) The MCF of cost 9
tree of G_1 connected

Fig. 10.3   An illustrative example of E-MCF method

new structure shown in Fig. 10.3b. Clearly, more links aggregate at some nodes. Figure 10.3c shows the final MCF getting by replacing each edge with the corresponding shortest path in G. This MCF for the same uncertain Multicast is of cost 9, which is much less than that of formed MCF in Fig. 10.2d.

Accordingly, we aim to introduce some shared nodes into the complete graph G_1. Thus, more edges with high aggregation probability will be added into the complete graph. Finally, the output MCF can save more links and reduce its total cost.

2. Identification of shared nodes

As mentioned above, shared nodes play a very important role in our E-MCF building method. However, it is unknown how to identify potentially shared nodes for an uncertain Multicast in the graph G. For this reason, we propose Algorithm 10.2 as follows.

In Algorithm 10.2, the function $ShortestPath$ calculates the shortest path between any two nodes. Note that the shortest paths from destinations to sources or those among destinations are involved, but that among sources are not involved since all sources will be finally disconnected. Accordingly, we extract intermediate nodes in each shortest path and count the times it is shared with other shortest paths via the function $IntermediateNode$. The function $IsConnected(u, v)$ is used to judge whether node u is connected with node v. An intermediate node is marked as a shared node if its counted frequency in the above process exceeds a preset threshold. This chapter sets this threshold as 3, while the more scientific and reasonable threshold is related to the specific configuration of the network and uncertain Multicast.

Algorithm 10.2 Identification of shared nodes

Require: An undirected graph $G = (V, E)$, the set of destinations D, and the set of sources S.
Ensure: A set of shared nodes P'.
1: $P' \leftarrow \emptyset, L \leftarrow \emptyset$;
2: **for** $\forall s_i \in S$ and $\forall d_j \in D$ **do**
3:     $L \leftarrow ShortestPath(s_i, d_j)$
4: **for** $\forall d_i \in D$ and $\forall d_j \in D$ and $d_i \neq d_j$ **do**
5:     $L \leftarrow ShortestPath(d_i, d_j)$
6: **for** $\forall l_i \in L$ **do**
7:     $Candidate \leftarrow IntermediateNode(l_i)$
8: **for** $\forall u \in Candidate$ **do**
9:     $NumberofNeighbor \leftarrow 0$
10:     **for** $\forall v \in Candidate, u \neq v$ **do**
11:         **if** $IsConnected(u, v)$ **then**
12:             $NumberofNeighbor \leftarrow NumberofNeighbor + 1$
13:     **if** $NumberofNeighbor > 3$ **then**
14:         $P' \leftarrow u$

As an example, we compare Fig. 10.2b with Fig. 10.3a. It is clear that nodes f and c are two branching points; hence, they are more possible to aggregate edges if they are included in the complete graph G_1. We can see shared nodes should be connected to many other nodes such that flow from such nodes will be immediately aggregated at a shared node.

3. **Analysis of E-MCF**

The major difference between P-MCF and E-MCF is the design of the complete graph G_1. As shown in Fig. 10.3, such a novel design is very effective to reduce the transmission cost of the same uncertain Multicast. Figure 10.3a shows the minimum spanning tree T of the new complete graph G_1 with the node set $\{1, 2, 3, 4, 5, 6, 7, s_1, s_2, s_3, f, c\}$. Notably, two shared nodes f and c are also included in G_1. To disconnect each pair of sources, two edges $\{c, 4\}$ and $\{s_3, 7\}$ are removed from T, and the resulted T_1 is shown in Fig. 10.3b. After replacing each edge in T_1 with the corresponding shortest path in G, we achieve the output MCF of cost 9, as shown in Fig. 10.3c.

Theorem 10.3 *Given any uncertain Multicast in a network $G = (V, E)$, E-MCF is more effective than P-MCF.*

Proof Recall that the P-MCF method first constructs the complete graph G_1 of all members of the uncertain Multicast and then derives a minimum spanning tree T_1 from G_1. We collect all intermediate nodes located on the shortest paths. The most frequent intermediate nodes among all shortest paths are selected as shared nodes. In the E-MCF method, those shared nodes will be added into G_1. Thus, a new complete graph G_2 and a minimum spanning tree T_2 are achieved. We will prove that T_2 might exhibit less total weight than T_1 due to the introduction of those shared nodes.

If adding a shared node v, some edges in T_1 might change according to two situations as follows. For each node u in T_1, T_2 will consider whether the new edge $e_{u,v}$ should be adopted.

1. For an edge $e_{u,w}$ in T_1, if the related shortest path in G traverses the shared node v, T_2 prefers to replace the edge $e_{u,w}$ with edges $e_{u,v}$ and $e_{v,w}$. This will not increase the total weight of T_2 since the weight of $e_{u,w}$ is equal to that of $e_{u,v}$ plus that of $e_{v,w}$. If other edges of node u in T_1, such as e_{u,w_1} and e_{u,w_2}, exhibit the same property of $e_{u,w}$, T_2 will update edge e_{u,w_1} with edges $e_{u,v}$ and e_{v,w_1}, and update e_{u,w_2} with edges $e_{u,v}$ and e_{v,w_2}. Thus, the common edge $e_{u,v}$ is shared repeatedly and can be aggregated. This fact will reduce the total weight of T_2.
2. Otherwise, let $e_{u,w}$ be the edge with the highest weight among all adjacent edges of u in T_1. If edge $e_{u,w}$ has higher weight than edge $e_{u,v}$, T_2 will adopt edge $e_{u,v}$ rather than edge $e_{u,w}$. The total weight of T_2 will be decreased by the weight of $e_{u,w}$ minus that of $e_{u,v}$. On the contrary, if the weight of edge $e_{u,w}$ is less than that of $e_{u,v}$, T_2 still utilizes edge $e_{u,w}$ even a shared node v is used.

In summary, the total weight of T_2 from E-MCF is less than that of T_1 from P-MCF. We have proved that P-MCF is $(2 + \varepsilon)$-approximation in Theorem 10.2. Thus, E-MCF will achieve a better approximation ratio than P-MCF.

We analyze the time complexity of the E-MCF method in the following way. In the beginning, we need to calculate $(|S| \cdot |D| + \frac{|D| \cdot |D-1|}{2})$ shortest paths from all destinations to all sources and among destinations. The time complexity of calculating one shortest path is $O(|V|^2)$ [18]; hence, the time complexity of finding the complete graph is $O((|S| \cdot |D| + \frac{|D| \cdot |D-1|}{2}) \times |V|^2)$. The time complexity of finding all shared nodes is $O(|V|)$. The time complexity of forming the minimum spanning tree T from G_1 is $O(|S| + |D| + |SharedNode|^2)$, where $|SharedNode|$ is the number of shared nodes. At last, removing some edges needs discovering the highest-weight edge in the path between any two sources in T. If each edge in T is checked, its complexity is $O(|S| + |D| + |SharedNode| - 1)$. In total, the complexity of our E-MCF method is $O((|S| \cdot |D| + \frac{|D| \cdot |D-1|}{2}) \times |V|^2)$.

10.5 Performance Evaluation

We start with the implementation method of uncertain Multicast in real networks. We then evaluate the performance of uncertain Multicast based on small-scale experiments. Finally, we conduct large-scale simulations to evaluate the uncertain Multicast under different network settings.

10.5.1 Implementation of the Uncertain Multicast in SDN Testbed

We execute the SMT method for traditional Multicast and our methods for uncertain Multicast in a real SDN network. 16 OpenFlow switches and the widely used controller RYU are adopted. We report the data plane and the control plane of our testbed as follows.

(1) Data Plane. Our testbed includes 16 ONetSwitch20 Openflow switches [20], each of which has four 1Gb data ports and one 1G management port. Each ONetSwitch20 switch consists of the Zynq SoC and other components so as to implement a line-rate packet switching. We select ONetSwitch20 as our data plane since it can flexibly support the management and customization of the flow table. Such ONetSwitch20 switches are connected randomly, and the available data ports of each switch are reserved for connecting computing nodes. Figure 10.2a plots the topology of our testbed, where all computing nodes are omitted. In this section, we vary the number of source and destination nodes, which connect to different switches. When we mention the source and destination nodes of a Multicast or an uncertain Multicast, we just replace each Multicast member with the related switch it appends in this section.

(2) Control Plane. We deploy the RYU controller [21] in the testbed since it is flexible to support our evaluation. We realize three Multicast applications, including the SMT, P-MCF, and E-MCF, on the RYU controller. The first task of the control plane is to generate and maintain the SMT for any given Multicast and the MCF for any given uncertain Multicast. The second task is to deploy the generated

Table 10.1 Multicast flow entry

Match	Actions
Multicast address	Output:port 1
	Output:port 2

SMT and MCF into the physical SDN network, according to the Openflow specification [22]. The packet forwarding procedure should be transparent such that switches need not know whether a packet belongs to a Multicast session or not. One simple method is to configure special addresses for Multicast sessions and configure Multicast flow entries, each of which consists of a set of output instructions. Once a Multicast flow entry in an involved switch matches packets with the address, the packet will be replicated and forwarded out from a given set of ports. Our flow table is shown in Table 10.1.

10.5.2   Evaluation Based on Small-Scale Experiments

In this section, we evaluate not only our P-MCF and E-MCF methods for uncertain Multicast but also the widely used SMT method on our testbed, whose topology is shown in Fig. 10.2a. The testbed keeps the topology unchanged but changes the configurations of uncertain Multicasts.

Given each setting of an uncertain Multicast, we first construct the desired forest using our P-MCF and E-MCF methods. At the same time, we derive a single-source Multicast by randomly picking a source from all potential sources. The SMT method establishes the minimum-cost tree for spanning the picked source and all destinations. We evaluate the resultant SMT or MCF using two performance metrics for any uncertain Multicast, including the total number of links and the flow completion time (FCT). The longest, shortest, and average FCT indicates the maximum, minimum, and average completion time among all flows in an uncertain Multicast, respectively. Note that each source delivers 10 MB of data towards related destinations.

1. **Impact of the number of sources**

We randomly fix 7 destination nodes while varying the number of sources from 2 to 6. Figure 10.4 reports the evaluation results about the three methods in terms of four metrics under the varied number of sources.

Figure 10.4a indicates that E-MCF always incurs a lower cost than the SMT when the number of sources increases from 2 to 6. This evidence proves the feasibility and effectiveness of the proposed uncertain Multicast in this chapter. Such benefit is more notable when an increasing number of sources are utilized. Additionally, the E-MCF also outperforms the P-MCF, irrespective of the number of sources.

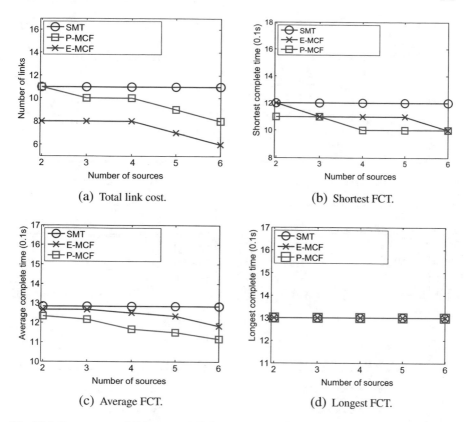

Fig. 10.4 Impact on total link cost and FCTs of uncertain Multicast when varying the numbers of sources from 2 to 6 in the testbed, while 7 destinations are fixed

Figure 10.4b, c show that the E-MCF and P-MCF of the uncertain Multicast outperform the SMT of the traditional Multicast in terms of the shortest and the average FCTs. Additionally, we find that the shortest and the average FCTs of P-MCF are less than that of E-MCF. The reason is that the P-MCF prefers to use the shortest path; hence, it contains the most amount of shortest paths from destinations to their nearest sources among three building algorithms. The longest FCT records the time duration until the last destination receives the data from sources; hence, it dominates the users' experience of the Multicast routing. In our small-scale testbed, the three methods exhibit the similar value of the longest FCT, as shown in Fig. 10.4d.

For the traditional SMT method, no matter an uncertain Multicast contains how many sources we consider a dedicated Multicast with one given source. Accordingly, the four performance metrics of the resultant SMT do not change with the increasing number of sources. When an uncertain Multicast employs more sources, E-MCF and P-MCF perform better than the SMT. The E-MCF is the desired one among the three methods since it incurs the least cost and lowers FCT.

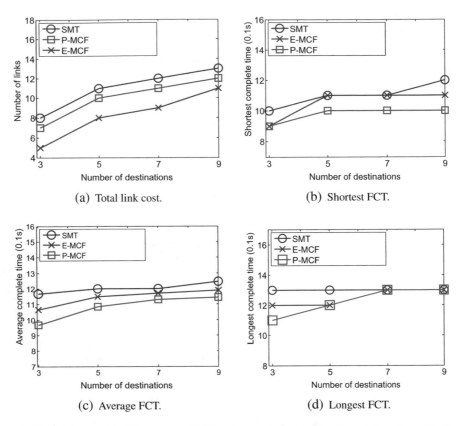

Fig. 10.5 Impact on total link cost and FCTs of uncertain multicast with varied number of destinations in the testbed, while two source nodes are fixed

2. Impact of the number of destinations

To evaluate the impact of the number of destinations, we randomly fix two source nodes and vary the number of destinations from 3 to 9. For the traditional SMT method, one source is randomly selected. Figure 10.5 reports the results about the three methods in terms of four metrics under the varied number of destination nodes.

Figure 10.5a shows that the total cost of the three methods increases along with increasing destinations. However, E-MCF and P-MCF always incur lower costs than SMT. Figure 10.5b, c, d report the shortest, average, and the largest FCTs under varied Multicast methods. We can see that the P-MCF method achieves the best performance in terms of the three FCT metrics, irrespective of the number of destinations. On the other hand, the E-MCF method incurs the lowest cost among the three methods, as shown in Fig. 10.5a, and achieves acceptable FCTs.

10.5.3 Evaluation Based on Large-Scale Simulations

Considering the limitations of the testbed scale, we carry out large-scale simulations to evaluate the performance of the three methods. Large-scale simulations differ from small-scale experiments in network topology, parameter configurations of uncertain Multicast, and evaluation metrics. More precisely, we conduct simulations in random networks, regular networks, and scale-free networks.

1. Simulation settings of large-scale networks

The above small-scale experimental results have demonstrated the benefits of our uncertain Multicast and proved that the E-MCF achieves the best performance than P-MCF and SMT. It is necessary to evaluate the performance of our methods under large-scale networks, where the uncertain Multicast involves more sources and destinations. Thus, we further conduct large-scale simulations under two performance metrics. They are the total link cost and the longest hop delay of each resultant Multicast routing structure. We use the largest hop length between destinations and sources to refer to the longest hop delay to ease the presentation.

Among the three networks, the regular network has the property that all nodes have the same degree. In the case of each network topology, we change the network scale from 1200 switches to 2800 switches, the number of sources from 3 to 21, and the number of destinations from 500 to 1000. The resultant network consists of 4800–10,000 hosts if each switch connects with four hosts. The two metrics are the total link cost and the longest hop delay between source and destination.

2. Evaluation under random networks

We first measure the performance of E-MCF, P-MCF, and SMT in large-scale random networks. We start with uncertain Multicasts with 10 sources and 300 destinations in varied scale random networks. Then the methods are evaluated under the case that the number of sources and destinations are varied.

Impact of the network scale. Figure 10.6 reports the evaluation results of the three methods when the number of switches increases in a random network. We can infer from Fig. 10.6a that the SMT of traditional Multicast causes the largest link cost among the three methods. The benefits of our P-MCF and E-MCF methods for the uncertain Multicast are more noticeable when the network scale becomes 2800. More precisely, our E-MCF method can save the total cost of traditional SMT by 39.23%. Figure 10.6b shows that our P-MCF is the best one among three methods in terms of the longest hop delay since any destination just selects the shortest path to its nearest source.

Although the P-MCF outperforms the E-MCF in terms of the longest hop delay, the E-MCF and SMT always occupy the least and most number of links, respectively. In summary, our E-MCF method always incurs the least cost than others, irrespective of the network scale.

Impact of the number of sources. In a random network with 2000 switches, we fix 300 destinations for an uncertain Multicast. As shown in Fig. 10.7a, with the

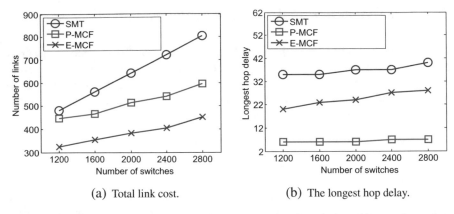

(a) Total link cost.   (b) The longest hop delay.

Fig. 10.6 Impact of varied number of switches on two metrics of uncertain multicast under random networks

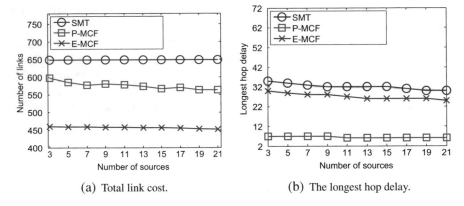

(a) Total link cost.   (b) The longest hop delay.

Fig. 10.7 Impact of varied number of sources on two metrics of uncertain multicast under random networks

increasing number of sources, the number of links based on the traditional SMT remains relatively stable because it just randomly utilizes one given source. Meanwhile, the E-MCF always incurs fewer links than the P-MCF and the SMT. Additionally, newly added sources will not considerably reduce the link costs when the number of sources increasing from 3 to 21. We can see from the evaluation results that three data copies (three sources) are somehow efficient and sufficient for uncertain Multicast in practice. Figure 10.7b shows that the P-MCF causes the least hop delay than the other two methods. It means that destinations can early receive data from sources in P-MCF than in E-MCF. In conclusion, E-MCF can not only reduce the hop delay of the SMT but also cause the least link cost among the three methods.

Impact of the number of destinations. In a random network with 2000 switches, we fix 10 sources for an uncertain Multicast. Figure 10.8 indicates that the addition of more destinations will naturally increase the link cost and the longest hop delay under

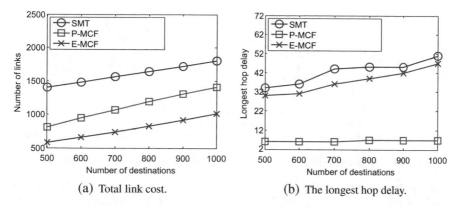

(a) Total link cost.   (b) The longest hop delay.

Fig. 10.8 Impact of varied number of destinations on two metrics of uncertain multicast under random networks

the three Multicast methods. Additionally, our two methods P-MCF and E-MCF, always outperform the traditional SMT method in terms of the two performance metrics. In summary, our E-MCF causes the lowest link cost while achieves the acceptable hop delay.

3. Evaluation under regular networks

To prove the effectiveness of uncertain Multicast in networks except for random networks, we evaluate the performance of the three methods in regular networks. The regular networks contain 1200–2800 switches. Under each set of network scale, we set the uncertain Multicast with 10 sources and 300 destinations. We then construct routing structures with P-MCF, E-MCF, and the traditional SMT, respectively. As shown in Fig. 10.9a, it is clear that the total link cost caused by the three methods increases along with the increasing network scale. Additionally, the E-MCF and P-MCF always cause less cost than the traditional SMT, irrespective of the network scale. Note that our E-MCF is the best one among those methods in terms of the total link cost.

Then, we select 300 destinations in a regular network with 2000 switches. In this setting, we change the number of sources from 3 to 21 and evaluate the performance of uncertain Multicast. We can see from Fig. 10.9b that the E-MCF with multiple sources is obviously better than the traditional SMT with only one source. The impact of more sources on the total link cost becomes weak when the number of sources exceeds 3. That is, the total link cost in the E-MCF slowly decreases when the number of sources increases. In this evaluation set, the E-MCF still occupies fewer links than the P-MCF and the traditional SMT.

Finally, we select 10 sources under the regular network, which includes 2000 switches. The number of destinations for uncertain Multicast changes from 500 to 1000. Figure 10.9c reports the evaluation results of three methods under the varied number of destinations. Similarly, our E-MCF always uses fewer links than the other two methods, irrespective of the number of destinations.

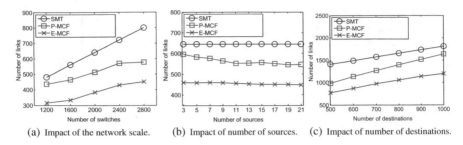

(a) Impact of the network scale.   (b) Impact of number of sources.   (c) Impact of number of destinations.

Fig. 10.9   The changing trends of the link cost under different environments in regular networks

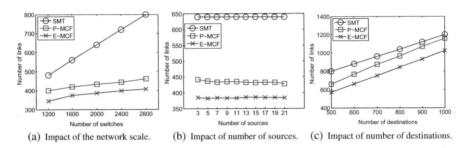

(a) Impact of the network scale.   (b) Impact of number of sources.   (c) Impact of number of destinations.

Fig. 10.10   The changing trends of the link cost under different environments in scale-free networks

4. **Evaluation under scale-free networks**

The Internet's topology conforms to scale-free networks to some extent. Thus, we further conduct simulations under scale-free networks in this section, evaluating the performance of the three methods for the same uncertain Multicast.

Firstly, we construct scale-free networks with different scales. The number of switches increases from 1200 to 2800. We randomly choose 10 sources and 300 destinations under each set of network scale. We then record the total link cost of our P-MCF, E-MCF, and the traditional SMT. As shown in Fig. 10.10a, it is clear that the total link cost of each method grows up along with the increasing network scale. It's noteworthy that the link cost of the SMT increases rapidly because it always utilizes a single source. With the SMT method, 800 links are occupied when the network involves 2800 switches. The increased link cost for our E-MCF and P-MCF is slow. It is clear that our E-MCF incurs the least number of links.

Secondly, in a scale-free network with 2000 switches and 300 destinations, we change the number of sources from 3 to 21. We can see from Fig. 10.10b that the SMT Multicast is not affected by the number of sources. The link cost of E-MCF and P-MCF becomes less when the number of available sources increases, and the E-MCF method can always incur the least number of links than the other two methods. Additionally, the impact of more sources on the E-MCF and P-MCF becomes weak when the number of sources exceeds 3.

Finally, in order to evaluate the impact of the number of destinations, we randomly choose 10 nodes as sources under scale-free networks with 2000 switches.

The number of destinations changes from 500 to 1000. As shown in Fig. 10.10c, three Multicasts need more links owing to the increase of receivers, and the E-MCF Multicast always occupies fewer links than the other two algorithms.

In conclusion, the P-MCF and E-MCF methods are adaptive to different network topologies, including random networks, regular networks, and scale-free networks, since they significantly reduce the link cost of uncertain Multicast. Additionally, the E-MCF method always incurs fewer links than the P-MCF method and the traditional SMT method, irrespective of the network topology, the network scale, the number of sources, and the number of destinations.

References

1. Mahimkar A A, Ge Z, Shaikh A, et al. Towards automated performance diagnosis in a large IPTV network[J]. ACM SIGCOMM Computer Communication Review, 2009, 39(4): 231–242.
2. Li D, Li Y, Wu J, et al. ESM: efficient and scalable data center multicast routing[J]. IEEE/ACM Transactions on Networking (TON), 2012, 20(3): 944–955.
3. Li D, Xu M, Liu Y, et al. Reliable multicast in data center networks[J]. Computers, IEEE Transactions on, 2014, 63(8): 2011–2024.
4. Guo D, Xie J, Zhou X, et al. Exploiting efficient and scalable shuffle transfers in future data center networks [J]. IEEE Transactions on Parallel & Distributed Systems, 2015, 26(4): 997–1009.
5. Kreutz D, Ramos F M V, Esteves Verissimo P, et al. Software-defined networking: A comprehensive survey [J]. Proceedings of the IEEE, 2015, 103(1): 14–76.
6. Robins G, Zelikovsky A. Tighter bounds for graph Steiner tree approximation [J]. SIAM Journal on Discrete Mathematics, 2005, 19(1): 122–134.
7. Shen S H, Huang L H, Yang D N, et al. Reliable Multicast Routing for Software-Defined Networks [J]. 2015: 181–189.
8. Chun B G, Wu P, Weatherspoon H, et al. Chunkcast: An anycast service for large content distribution [C]. In Proc. of 5th International IPTPS, Santa Barbara, CA, USA, 2006.
9. Kou L, Markowsky G, Berman L. A fast algorithm for Steiner trees [J]. Acta informatica, 1981, 15(2): 141–145.
10. Robins G, Zelikovsky A. Improved Steiner tree approximation in graphs [C]. In Proc. of 11th ACM-SIAM SODA, 2000: 770–779.
11. Karpinski M, Zelikovsky A. New approximation algorithms for the steiner tree problems [J]. Journal of Combinatorial Optimization, 1997, 1(1): 47–65.
12. Huang L H, Hung H J, Lin C C, et al. Scalable Steiner Tree for Multicast Communications in Software-Defined Networking [J]. arXiv preprint arXiv, 2014.
13. Zhang S, Zhang Q, Bannazadeh H, et al. Routing algorithms for network function virtualization enabled multicast topology on SDN [J]. IEEE Transactions on Network and Service Management, 2015, 12(4): 580–594.
14. Gu W, Zhang X, Gong B, et al. A survey of multicast in software-defined networking [C]. In Proc. of 5th ICIMM, Hohhot, China, 2015.
15. Chen Y R, Radhakrishnan S, Dhall S, et al. On multi-stream multi-source multicast routing [J]. Computer Networks, 2013, 57(13): 2916–2930.
16. Robins G, Zelikovsky A. Minimum steiner tree construction [J]. The Handbook of Algorithms for VLSI Phys. Design Automation, 2009: 487–508.
17. Zheng X, Cho C, Xia Y. Content distribution by multiple multicast trees and intersession cooperation: Optimal algorithms and approximations [J]. Computer Networks, 2015, 83: 5857–5862.

18. Du D, Ko K I, Hu X. Design and Analysis of Approximation Algorithms [J]. Higher Education Press, 2011, 62.
19. Zhong C, Malinen M, Miao D, et al. A fast minimum spanning tree algorithm based on K-means [J]. Information Sciences, 2015, 295: 1–17.
20. ONetSwitch [EB/OL]. [2016-01-18]. http://www.meshsr.com/product/onetswitch20.
21. RYU Controller Tutorial [EB/OL]. [2016-01-18]. http://sdnhub.org/tutorials/ryu/.
22. McKeown N, Anderson T, Balakrishnan H, et al. OpenFlow: enabling innovation in campus networks [J]. ACM SIGCOMM Computer Communication Review, 2008, 38(2): 69–74.

Printed in the United States
by Baker & Taylor Publisher Services